AN APPETITE FOR WONDER

An Appetite for Wonder

The Making of a Scientist

A Memoir

Richard Dawkins

An Imprint of HarperCollins*Publishers*

HarperCollins books may be purchased for educational, business, or sales promotional use. For information please e-mail the Special Markets Department at SPsales@harpercollins.com.

FIRST U.S. EDITION

Originally published in Great Britain in 2013 by Bantam Press, an imprint of Transworld Publishers.

Diagrams by Patrick Mulrey

Library of Congress Cataloging-in-Publication Data has been applied for.

ISBN 978-0-06-222579-5 (hardcover)
ISBN 978-0-06-228715-1 (international edition)
ISBN 978-0-06-231580-9 (signed edition)

13 14 15 16 17   OV/RRD   10 9 8 7 6 5 4 3 2 1

To my mother and my sister, who shared the years with me,
and in memory of my father, missed by all.

CONTENTS

The Over Norton
connection:

James Dawkins
MP (1696–1766)
uncle to ----------- Henry Dawkins MP *m* Lady Juliana   General Sir Henry
                      (1728–1814)   Colyear   Clinton (1738–95)

                           Henry Dawkins *m* Augusta Clinton
                           (1765–1852)

Henry Dawkins   Clinton George Augustus
                                 'Cannonball' Dawkins (1808–71)

                      Francis Henry   (*later Sir*) Clinton
                      Dawkins   Edward Dawkins

Col. William
Gregory Dawkins   Dr James Wearne
(1825–1914)
great-uncle to   Dr Walter Wearne

                      Ethel   Connie *m* Alan Wilfr
                      Wearne   Wearne   'Bill' Ladn

Major Hereward
Dawkins   (Clinton) *m* Jean Mary
3rd cousin to ------------------------ John Dawkins   Vyvyan Ladner

                           (CLINTON)   *m* (1) Marian Stamp
                           RICHARD
                           DAWKINS   *m* (2) Eve Barham (*dec'd*)

                                        m (3) (Sarah)
                                              Lalla Ward

                           Juliet Dawkins

> *The stuff of life to knit me*
> *Blew hither; here am I.*
>
> A. E. Housman, A Shropshire Lad

Revd William Smythies
(159?–1653)

....

7 generations
of Anglican vicars

....

Arthur Smythies

Evelyn *m* Olive
Smythies | Cripps

Bernard 'Bunny'
Smythies

Bertram 'Billy' Smythies   John Smythies

Yorick Smythies

Clinton George Evelyn Dawkins *m* Enid Smythies

Diana *m* (1) Bob Keddie
Ladner

m (2) Arthur Francis
'Bill' Dawkins

(Henry) Colyear
Dawkins

Sarah Dawkins *m* Michael Kettlewell

Penny
Keddie

Thomas
Dawkins

GENES AND
PITH HELMETS

'GLAD to know you, Clint.' The friendly passport controller was not to know that British people are sometimes given a family name first, followed by the name their parents wanted them to use. I was always to be Richard, just as my father was always John. Our first name of Clinton was something we forgot about, as our parents had intended. To me, it has been no more than a niggling irritation which I would have been happier without (notwithstanding the serendipitous realization that it gives me the same initials as Charles Robert Darwin). But alas, nobody anticipated the United States Department of Homeland Security. Not content with scanning our shoes and rationing our toothpaste, they decreed that anyone entering America must travel under his first name, exactly as written in his passport. So I had to forgo my lifelong identity as Richard and rebrand myself Clinton R. Dawkins when booking tickets to the States – and, of course, when filling in those important forms: the ones that require you explicitly to deny that you are entering the USA in order to overthrow the constitution by force of arms. ('Sole purpose of visit' was the British broadcaster Gilbert Harding's response to that; nowadays such levity will see you banged up.)

Clinton Richard Dawkins, then, is the name on my birth

certificate and passport, and my father was Clinton John. As it happened, he was not the only C. Dawkins whose name appeared in *The Times* as the father of a boy born in the Eskotene Nursing Home, Nairobi, in March 1941. The other was the Reverend Cuthbert Dawkins, Anglican missionary and no relation. My bemused mother received a shower of congratulations from bishops and clerics in England, unknown to her but kindly calling down God's blessings upon her newborn son. We cannot know whether the misdirected benedictions intended for Cuthbert's son had any improving effect on me, but he became a missionary like his father and I became a biologist like mine. To this day my mother jokes that I might be the wrong one. I am happy to say that more than just my physical resemblance to my father reassures me that I am not a changeling, and was never destined for the church.

Clinton first became a Dawkins family name when my great-great-great-grandfather Henry Dawkins (1765–1852) married Augusta, daughter of General Sir Henry Clinton (1738–95), who, as Commander-in-Chief of British forces from 1778 to 1782, was partly responsible for losing the American War of Independence. The circumstances of the marriage make the commandeering of his name by the Dawkins family seem a bit cheeky. The following extract is from a history of Great Portland Street, where General Clinton lived.

> In 1788 his daughter eloped from this street in a hackney-coach with Mr Dawkins, who eluded pursuit by posting half a dozen other hackney-coaches at the corners of the street leading into Portland Place, with directions to drive off as rapidly as possible, each in a different direction . . .[1]

---

[1] H. B Wheatley and P. Cunningham, *London Past and Present* (London, Murray, 1891), vol. 1, p. 109.

I wish I could claim this ornament of the family escutcheon as the inspiration for Stephen Leacock's Lord Ronald, who '. . . flung himself upon his horse and rode madly off in all directions'. I'd also like to think that I inherited some of Henry Dawkins's resourcefulness, not to mention his ardour. This is unlikely, however, as only one 32nd part of my genome is derived from him. One 64th part is from General Clinton himself, and I have never shown any military leanings. *Tess of the D'Urbervilles* and *The Hound of the Baskervilles* are not the only works of fiction that invoke hereditary 'throwbacks' to distant ancestors, forgetting that the proportion of genes shared is halved with every generation and therefore dies away exponentially – or it would if it were not for cousin-marriage, which becomes ever more frequent the more distant the cousinship, so that we are all more or less distant cousins of each other.

It is a remarkable fact, which you can prove to yourself without leaving your armchair, that if you go back far enough in a time machine, any individual you meet who has any living human descendants at all must be an ancestor of everybody living. When your time machine has travelled sufficiently far into the past, everybody you meet is an ancestor either of everybody alive in 2013 or of nobody. By the method of *reductio ad absurdum* beloved of mathematicians, you can see that this has to be true of our fishy ancestors of the Devonian era (my fish has to be the same as your fish, because the absurd alternative is that your fish's descendants and my fish's descendants stayed chastely separate from each other for 300 million years yet are still capable of interbreeding today). The only question is how far back you have to go to apply that argument. Clearly not as far as our fishy forebears, but how far? Well, hurdling swiftly over the detailed calculation, I can tell you that if the Queen is descended from William the Conqueror, you quite probably are too (and – give or take the odd illegitimacy – I know I am, as does almost everybody with a recorded pedigree).

Henry and Augusta's son, Clinton George Augustus Dawkins (1808–71) was one of the few Dawkinses actually to use the name Clinton. If he inherited any of his father's ardour he nearly lost it in 1849 during an Austrian bombardment of Venice, where he was the British consul. I have a cannonball in my possession, sitting on a plinth bearing an inscription on a brass plate. I don't know whose is the authorial voice and I don't know how reliable it is, but, for what it is worth, here is my translation (from French, then the language of diplomacy):

> One night when he was in bed, a cannonball penetrated the bed covers and passed between his legs, but happily did him no more than superficial damage. I first took this to be a tall story, until I learned for certain that it was based on the exact truth. His Swiss colleague met him later in the funeral procession of the American consul and, when asked about it, he laughingly confirmed the facts and told him it was precisely for this reason he was limping.

This narrow escape of my ancestor's vital parts took place before he was to put them to use, and it is tempting to attribute my own existence to a stroke of ballistic luck. A few inches closer to the fork of Shakespeare's radish and . . . But actually my existence, and yours, and the postman's, hangs from a far narrower thread of luck than that. We owe it to the precise timing and placing of everything that ever happened since the universe began. The incident of the cannonball is only a dramatic example of a much more general phenomenon. As I have put it before, if the second dinosaur to the left of the tall cycad tree had not happened to sneeze and thereby fail to catch the tiny, shrew-like ancestor of all the mammals, we would none of us be here. We all can regard ourselves as exquisitely improbable. But here, in a triumph of hindsight, we are.

C. G. A. ('Cannonball') Dawkins's son Clinton (later Sir Clinton)

Edward Dawkins (1859–1905) was one of many Dawkinses to attend Balliol College, Oxford. He was there at the right time to be immortalized in the Balliol Rhymes, originally published as a broadsheet called *The Masque of Balliol* in 1881. In the spring term of that year, seven undergraduates composed and printed scurrilous rhymes about personalities of the college. Most famous is the verse that celebrates Balliol's great Master, Benjamin Jowett, composed by H. C. Beeching, later Dean of Norwich Cathedral:

> First come I, my name is Jowett.
> There's no knowledge but I know it.
> I am Master of this College,
> What I don't know isn't knowledge.

Less witty, but intriguing to me, is the rhyme on Clinton Edward Dawkins:

> Positivists ever talk in s-
> Uch an epic style as Dawkins;
> God is naught and Man is all,
> Spell him with a capital.

Freethinkers were much less common in Victorian times, and I wish I had met great-great-uncle Clinton (as a child I did meet two of his younger sisters in advanced old age, one of whom had two maids called – I found the surname convention weird – Johnson and Harris). And what should we make of that 'epic style'?

I believe Sir Clinton later paid for my grandfather, his nephew Clinton George Evelyn Dawkins, to go to Balliol, where he seems to have done little but row. There is a photograph (reproduced in the picture section) of my grandfather preparing for action on the river that is wonderfully evocative of Edwardian high summer in Oxford.

It could be a scene from Max Beerbohm's *Zuleika Dobson*. The behatted guests are standing on the college barge, the floating boathouse which all the college rowing clubs maintained until living memory. Today, alas, they have been replaced by serviceable brick boathouses on the shore. (One or two of the barges are still afloat – or at least aground – as houseboats, having been towed to watery resting places amid moorhens and grebes in the backwaters and rivers around Oxford.) The resemblance between Grandfather and two of his sons, my father and my Uncle Colyear, is unmistakable. Family resemblances fascinate me, although they die away rapidly as the generations march on.

Grandfather was devoted to Balliol and contrived to stay there far beyond the normally allotted span of an undergraduate – solely, I suspect, in order to carry on rowing. When I used to visit him in old age, the college was his main topic of conversation, and he repeatedly wanted to know whether we still used (I repeatedly had to tell him we didn't) the same Edwardian slang: 'Mugger' for Master; 'wagger pagger' for wastepaper basket; Maggers' Memogger for the Martyrs' Memorial, the landmark cross outside Balliol that commemorates the three Anglican bishops who were burned alive in Oxford in 1555 for their attachment to the wrong flavour of Christianity.

One of my last memories of Grandfather Dawkins was of delivering him to his final Balliol gaudy (reunion dinner for former members, where each year a different age cohort is entertained). Surrounded by old comrades pushing Zimmer frames ('walkers') and festooned with ear trumpets and pince-nez, he was recognized by one of them who indulged the obvious sarcasm: 'Hello, Dawkins, you still rowing for Leander?' I left him looking a trifle forlorn among the boys of the old brigade, some of whom must surely have fought in the Boer War and were, therefore, rightful dedicatees of Hilaire Belloc's famous poem 'To the Balliol Men Still in Africa':

Years ago, when I was at Balliol,
Balliol men – and I was one –
Swam together in winter rivers,
Wrestled together under the sun.
And still in the heart of us, Balliol, Balliol,
Loved already, but hardly known,
Welded us each of us into the others:
Called a levy and chose her own.
Here is a House that armours a man
With the eyes of a boy and the heart of a ranger
And a laughing way in the teeth of the world
And a holy hunger and thirst for danger:

Balliol made me, Balliol fed me,
Whatever I had she gave me again:
And the best of Balliol loved and led me.
God be with you, Balliol men.

With difficulty I read this at my father's funeral in 2011, and then again in 2012 when I gave a eulogy for Christopher Hitchens, another Balliol man, at the Global Atheist Convention in Melbourne. With difficulty because, even on happier occasions, I become tearful with embarrassing ease when reciting loved poetry, and this particular poem by Belloc is one of the worst offenders.

After leaving Balliol, Grandfather made his career, like so many of my family, in the Colonial Service. He became Conservator of Forests in his district of Burma, where he spent much time in remote corners of the hardwood forests, supervising the heavy work of the highly trained elephant lumberjacks. He was up-country among the teak trees when the news reached him – I like to fancy by hand of runner with cleft stick – of the birth, in 1921, of his youngest son Colyear (named after Lady Juliana Colyear, mother of the enterprising Henry who eloped with Augusta Clinton). He was so excited that, without waiting for other transport to be available, he bicycled 50 miles to be at his wife Enid's bedside, where he proudly opined that the new boy had the 'Dawkins nose'. Evolutionary psychologists have noted the particular eagerness with which new babies are scanned for resemblances to their paternal, as opposed to maternal relatives – for the obvious reason that it is harder to be confident of paternity than maternity.

Colyear was the youngest and John, my father, the eldest of three brothers, all of whom were born in Burma to be carried around the jungle in Moses baskets slung from poles by trusty bearers, and all of whom eventually followed their father into the Colonial Service, but in three different parts of Africa: John in Nyasaland (now Malawi), the middle brother, Bill, in Sierra Leone, and Colyear in Uganda. Bill was christened Arthur Francis after his two grandfathers, but was always called Bill for a childhood resemblance to Lewis Carroll's Bill the Lizard. John and Colyear looked alike as young men, to the extent that John was once stopped in the street and asked: 'Are you you or your brother?' (That story is true, which is perhaps more than can be said of the famous legend that W. A. Spooner, the only Warden (head) of my present Oxford college to qualify for an 'ism', once greeted a young man in the quad with the question: 'Let me see, I never can remember, was it you or your brother was killed in the war?') As they aged, Bill and Colyear grew more alike (and like their father) and John less so, to my eyes. It often happens that family resemblances appear and disappear at different stages during a life history, which is one reason I find them fascinating. It is easy to forget that genes continue to exert their effects throughout life, not just during embryonic development.

There was no sister, to the regret of my grandparents, who had intended that their youngest would be Juliana but had to settle for her noble surname instead. All three brothers were talented. Colyear was the cleverest academically, and Bill the most athletic: I was proud to see his name on the roll of honour at the school I attended later, as holder of the school record for the hundred yards sprint – an ability which doubtless served him well at rugby when he scored a dashing touchdown for the Army against Great Britain early in the Second World War. I share none of Bill's athleticism, but I like to think that I learned how to think about science from my father, and how to explain it from my Uncle Colyear. Colyear became an Oxford don after leaving Uganda and was widely revered as a brilliant teacher of statistics, a notoriously difficult subject to convey to biologists. He died too young, and I dedicated one of my books, *River Out of Eden*, to him in the following terms:

> To the memory of Henry Colyear Dawkins (1921–1992), Fellow of St John's College, Oxford: a master of the art of making things clear.

The brothers died in reverse order of age and I sadly miss them all. I spoke the eulogy at the funeral of Bill, my godfather and uncle, when he died at the age of 93 in 2009.[1] I tried to convey the idea that, although there was much that was bad in the British Colonial Service, the best was very good indeed; and Bill, like his two brothers, and like Dick Kettlewell whom I'll mention later,[2] was of the best.

If the three brothers could be said to have followed their father into the Colonial Service, they had a similar heritage on their mother's side too. Their maternal grandfather, Arthur Smythies, was Chief Conservator of Forests in his district of India; his son Evelyn became Chief Conservator of Forests in Nepal. It was my Dawkins grandfather's friendship with Evelyn, forged while both were

---

[1] See web appendix: www.richarddawkins.net/afw.
[2] And whose obituary I wrote: see web appendix.

reading forestry at Oxford, that led to his meeting and marrying Evelyn's sister Enid, my grandmother. Evelyn was the author of a noted book on *India's Forest Wealth* (1925) as well as various standard works on philately. His wife Olive, I am sorry to say, was fond of shooting tigers and published a book called *Tiger Lady*. There is a picture of her standing on a tiger and under a solar topee, with her husband proudly patting her on the shoulder, captioned: 'Well done, little woman.' I don't think she would have been my type.

Olive and Evelyn's eldest son, my father's taciturn first cousin Bertram ('Billy') Smythies, was also in the forest service, in Burma and later Sarawak: he wrote the standard works *Birds of Burma* and *Birds of Borneo*. The latter became a kind of bible to the (not at all taciturn) travel writer Redmond O'Hanlon, on his hilarious journey *Into the Heart of Borneo* with the poet James Fenton.

Bertram's younger brother John Smythies departed from family tradition and became a distinguished neuroscientist and authority on schizophrenia and psychedelic drugs, living in California, where he is credited with inspiring Aldous Huxley to take mescaline and cleanse his 'doors of perception'. I recently asked his advice on whether to accept the kind offer of a friend to mentor me through an LSD trip. He advised against. Yorick Smythies, another first cousin of my father, was a devoted amanuensis of the philosopher Wittgenstein.[1] Peter Conradi, in his biography of the novelist Iris Murdoch, identifies Yorick as the 'holy fool' upon whom she based one of the characters in *Under the Net*, Hugo Belfounder. I must say it is hard to see the resemblance.

> Yorick wished to become a bus conductor but, [Iris Murdoch] noted, was the only person in the history of the bus company to fail the theory test ... During his single driving-lesson

[1] http://wab.uib.no/ojs/agora-alws/article/view/1263/977

the instructor left the car as Yorick drove on and off the
pavement.

Having failed to make the grade as a bus conductor, and
dissuaded by Wittgenstein (along with most of his other pupils)
from a career in philosophy, Yorick worked as a librarian in the
Oxford forestry department, which may have been his only
connection with the family tradition. He had eccentric habits, took
to snuff and Roman Catholicism, and died tragically.

Arthur Smythies, grandfather to the Dawkins and Smythies
cousins, seems to have been the first in my family to enter Imperial
service. His paternal ancestors for seven unbroken generations back
to his great-great-great-great-great-grandfather (the Reverend
William Smythies, born in the 1590s) were Anglican clergy to a man.
I suppose it is not unlikely that, had I lived in any of their centuries,
I might have been a clergyman too. I have always been interested in
the deep questions of existence, the questions that religion aspires
(and fails) to answer, but I have been fortunate to live in a time when
such questions are given scientific rather than supernatural answers.
Indeed, my interest in biology has been largely driven by questions
about origins and the nature of life, rather than – as is the case for
most young biologists I have taught – by a love of natural history. I
might even be said to have let down the family tradition of devotion
to outdoor pursuits and field natural history. In a brief previous
memoir published in an anthology of autobiographical chapters by
ethologists, I wrote:

> I should have been a child naturalist. I had every advantage: not
> only the perfect early environment of tropical Africa but what
> should have been the perfect genes to slot into it. For generations,
> sun-browned Dawkins legs have been striding in khaki shorts
> through the jungles of Empire. Like my father and his two

younger brothers, I was all but born with a pith helmet on my head.[1]

Indeed, my Uncle Colyear was later to say, on seeing me in shorts for the first time (he habitually wore them himself, held up by two belts): 'Good God, you've got authentic Dawkins knees.' I went on to write of my Uncle Colyear that the worst thing he could say of a young man was:

'Never been in a youth hostel in his life'; a stricture, which, I am sorry to say, describes me to this day. My young self seemed to let down the traditions of the family.

I received every encouragement from my parents, both of whom knew all the wildflowers you might encounter on a Cornish cliff or an Alpine meadow, and my father amused my sister and me by throwing in the Latin names for good measure (children love the sound of words even if they don't know their meanings). Soon after arriving in England, I was mortified when my tall, handsome grandfather, by now retired from the Burma forests, pointed to a blue tit outside the window and asked me if I knew what it was. I didn't and miserably stammered, 'Is it a chaffinch?' Grandfather was scandalized. In the Dawkins family, such ignorance was tantamount to not having heard of Shakespeare: 'Good God, John' – I have never forgotten his words, nor my father's loyal exculpation – 'is that *possible?*'

To be fair to my young self, I had only just set foot in England, and neither blue tits nor chaffinches occur in east Africa. But in any case I learned late to love watching wild creatures, and I have never

---

[1] 'Growing up in ethology', ch. 8 in L. Drickamer and D. Dewsbury, eds, *Leaders in Animal Behavior* (Cambridge, Cambridge University Press, 2010).

been such an outdoor person as either my father or my grandfather. Instead:

> I became a secret reader. In the holidays from boarding school, I would sneak up to my bedroom with a book: a guilty truant from the fresh air and the virtuous outdoors. And when I started learning biology properly at school, it was still bookish pursuits that held me. I was drawn to questions that grown-ups would have called philosophical. What is the meaning of life? Why are we here? How did it all start?

My mother's family came from Cornwall. Her mother, Connie Wearne, was the daughter and grand-daughter of Helston doctors (as a child I imagined them both as Dr Livesey in *Treasure Island*). She was herself fiercely Cornish, referring to the English as 'foreigners'. She regretted having been born too late to speak the now extinct Cornish language, but she told me that when she was a girl the old Mullion fishermen could understand the Breton fishermen 'who came to pinch our crabs'. Of the Brythonic languages, Welsh (alive), Breton (dying) and Cornish (dead), Breton and Cornish are sister species on the language family tree. A number of Cornish words survive in the Cornish dialect of English, for example *quilkin* for frog, and my grandmother could do the dialect well. We, her grandchildren, repeatedly persuaded her to recite a lovely rhyme about a boy who 'clunked a bully' (swallowed a plumstone). I even recorded one of these recitations, and sadly regret that I have lost the tape. Much later, Google helped me to track down the words,[1] and I can still hear her squeaky voice saying them in my head.

---

[1] From *Randigal Rhymes*, ed. Joseph Thomas (Penzance, F. Rodda, 1895).

There was an awful pop and towse[1] just now down by the hully,[2]
For that there boy of Ben Trembaa's, aw went and clunked[3] a bully,[4]
Aw ded'n clunk en fitty,[5] for aw sticked right in his uzzle,[6]
And how to get en out again, I tell ee 'twas a puzzle,
For aw got chucked,[7] and gasped, and urged,[8] and rolled his eyes,
   and glazed;
Aw guggled, and aw stank'd[9] about as ef aw had gone mazed.[10]

Ould Mally Gendall was the fust that came to his relief,–
Like Jimmy Eellis 'mong the cats,[11] she's always head and chief;
She scruffed 'n by the cob,[12] and then, before aw could say 'No,'
She fooched her finger down his throat as fur as it would go,
But aw soon catched en 'tween his teeth, and chawed en all the
   while,
Till she screeched like a whitneck[13]–you could hear her 'most a mile;

And nobody could help the boy, all were in such a fright,
And one said: 'Turn a crickmole,[14] son; 'tes sure to put ee right;'
And some ran for stillwaters,[15] and uncle Tommy Wilkin
Began a randigal[16] about a boy that clunked a quilkin;[17]
Some shaked their heads, and gravely said: "Twas always clear to them

---

[1] Fuss.
[2] Store for live bait.
[3] Swallowed.
[4] Pebble, though my grandmother translated it as plumstone, which makes more sense.
[5] Properly.
[6] Throat.
[7] Choked.
[8] Retched.
[9] Stamped.
[10] Mad.
[11] Local proverb.
[12] Forelock.
[13] Stoat, weasel.
[14] Somersault.
[15] Medicine distilled from peppermint.
[16] Nonsensical story.
[17] Swallowed a frog.

That boy'd end badly, for aw was a most anointed lem,[1]
For aw would minchey,[2] play at feaps,[3] or prall[4] a dog or cat,
Or strub[5] a nest, unhang a gate, or anything like that.'

Just then Great Jem stroathed[6] down the lane, and shouted out so
    bold:
'You're like the Ruan Vean men, soase, don't knaw and waant be
    told;'
Aw staved right in amongst them, and aw fetched that boy a clout,
Just down below the nuddick,[7] and aw scat the bully out;
That there's the boy that's standing where the keggas are in
    blowth:[8]
Blest! If aw haven't got another bully in his mouth!'

I am fascinated by the evolution of language, and how local versions diverge to become dialects like Cornish English and Geordie and then imperceptibly diverge further to become mutually unintelligible but obviously related languages like German and Dutch. The analogy to genetic evolution is close enough to be illuminating and misleading at the same time. When populations diverge to become species, the time of separation is defined as the moment when they can no longer interbreed. I suggest that two dialects should be deemed to reach the status of separate languages when they have diverged to an analogously critical point: the point where, if a native speaker of one attempts to speak the other it is taken as a compliment rather than as an insult. If I went into a

---

[1] Mischievous imp.
[2] Truant.
[3] Pitch and toss.
[4] Tie a tin can or something to an animal's tail.
[5] Rob.
[6] Briskly strode.
[7] Back of the head.
[8] Cow parsleys are in bloom.

Penzance pub and attempted to speak the Cornish dialect of English I'd be asking for trouble, because I'd be heard as mockingly imitating. But if I go to Germany and attempt to speak German, people are delighted. German and English have had enough time to diverge. If I am right, there should be examples – maybe in Scandinavia? – where dialects are on the cusp of becoming separate languages. On a recent lecturing trip to Stockholm I was a guest on a television talk show which was aired in both Sweden and Norway. The host was Norwegian, as were some of the guests, and I was told that it didn't matter which of the two languages was spoken: audiences on both sides of the border effortlessly understand both. Danish, on the other hand, is difficult for most Swedes to understand. My theory would predict that a Swede visiting Norway would probably be advised not to attempt to speak Norwegian for fear of being thought insulting. But a Swede visiting Denmark would probably be popular if she attempted to speak Danish.[1]

When my great-grandfather Dr Walter Wearne died, his widow moved out of Helston and built a house overlooking Mullion Cove on the west side of the Lizard peninsula, which has remained in the family ever since. A lovely cliff walk among the sea pinks from Mullion Cove takes you to Poldhu, site of Guglielmo Marconi's radio station from which the first ever transatlantic radio transmission was sent in 1901. It consisted of the letter 's' in Morse code, repeated over and over. How could they be so dull, on such a momentous occasion, as to say nothing more imaginative than s s s s s?

My maternal grandfather, Alan Wilfred 'Bill' Ladner, was Cornish too, a radio engineer employed by the Marconi company. He joined too late to be involved in the 1901 transmission but he was sent to work at the same radio station at Poldhu around 1913,

---

[1] I've consulted an expert on Scandinavian languages, Professor Björn Melander, and he agreed with my theory of 'insult or flattery' but added that there are, inevitably, complications of context.

shortly before the First World War. When the Poldhu Wireless Station was finally dismantled in 1933, my grandmother's elder sister Ethel (known simply as 'Aunt' to my mother, although she wasn't her only aunt) was able to acquire some large slate slabs that had been used as instrument panels, with holes drilled in them in patterns that traced out their use – fossils of a bygone technology. These slates now pave the garden of the family house at Mullion (see the picture section), where they inspired me, as a boy, to admiration of my grandfather's honourable profession of engineer – honoured less in Britain than in many other countries, which may go some way towards explaining my country's sad decline from a once great manufacturing power to the indignity of being a provider of (often, as we now sadly know, rather dodgy) 'financial services'.

Before Marconi's historic transmission, the distance across which radio signals could be received was believed to be limited by the curvature of the Earth. How could waves that travelled in a straight line be picked up beyond the horizon? The solution proved to be that waves could bounce off the Heaviside Layer in the upper atmosphere (and modern radio signals, of course, bounce off artificial satellites instead). I am proud that my grandfather's book, *Short Wave Wireless Communication*, went through many editions from the 1930s to the early 1950s as the standard textbook on the subject, until it was eventually superseded around the time when valves[1] were replaced by transistors.

That book was always legendary in the family for its in-comprehensibility, but I have just read the first two pages and find myself delighted by its lucidity.

The ideal transmitter would produce an electrical signal which was a faithful copy of the impressed signal and would transmit

---

[1] 'Vacuum tubes' in American English.

this to the connecting link with perfect constancy and in such a manner that no interference was caused to other channels. The ideal connecting link would transmit the electric impulses through or over it without distorting them, without attenuation, and would collect no 'noise' on the way from extraneous electrical disturbances of whatever kind. The ideal receiver would pick up the required electrical impulses despatched through the connecting link by the transmitter of the channel and transform them with perfect faithfulness into the required form for visual or audible observation . . . As it is very unlikely that the ideal channel will ever be developed, we must consider in what directions we would prefer to compromise.

Sorry, Grandfather; sorry I was put off reading your book while you were still around to talk about it – and when I was old enough to understand it but was put off even trying. And you were put off by family pressure, put off ever divulging the rich store of knowledge that must have been there still in your clever old brain. 'No, I don't know anything about wireless,' you would mutter to any overture, and then resume your near ceaseless whistling of light opera under your breath. I would love to talk to you now about Claude Shannon and information theory. I would love to show you how just the same principles govern communication between bees, between birds, and indeed between neurones in the brain. I would love you to teach me about Fourier transforms and reminisce about Professor Silvanus Thompson, author of *Calculus Made Easy* ('What one fool can do, another can'). So many missed opportunities, gone for ever. How could I have been so short-sighted, so dull? Sorry, shade of Alan Wilfred Ladner, Marconiman and beloved grandfather.

It was my Uncle Colyear rather than my Grandfather Ladner who prompted me to try to build radios in my teens. He gave me a book by F. J. Camm, from which I took the plans to build first a crystal set (which just faintly worked) and then a one-valve set – with a large, bright red valve – which worked slightly better but still needed headphones rather than a loudspeaker. It was unbelievably badly made. Far from arranging the wires tidily, I took delight in the fact that it didn't matter how untidy were the pathways they took, stapled down on a wooden chassis, so long as each wire ended up in the right place. I won't say I went out of my way to make the course of each wire untidy, but I certainly was fascinated by the mismatch between the topology of the wires, which really mattered, and their physical layout, which didn't. The contrast with a modern integrated circuit is staggering. Many years later, when I gave the Royal Institution Christmas Lectures to children of about the same age as I was when I made my one-valve set, I borrowed the hugely magnified layout diagram of an integrated circuit from a modern computer company to show them. I hope my young auditors were awestruck and a bit bewildered by it. Experimental embryologists have shown that growing nerve cells often sniff out their correct end organs in something like the way I built my one-valve set, rather than by following an orderly plan like an integrated circuit.

Back to Cornwall before the First World War. It was my great-grandmother's habit to invite the lonely young engineers from the clifftop radio station to tea at Mullion, and that was how my grandparents met. They became engaged, but then the war broke out. Bill Ladner's skills as a radio engineer were in demand, and he was sent by the Royal Navy as a smart young officer to the southern tip of what was then Ceylon to build a radio station at that strategically vital staging post in the Empire's shipping lanes.

Connie followed him out in 1915, where she stayed in a local vicarage, from which they were married. My mother, Jean Mary Vyvyan Ladner, was born in Colombo in 1916.

In 1919, the war over, Bill Ladner brought his family back to

England: not to Cornwall in the far west of the country but to Essex in the far east, where the Marconi company had its headquarters in Chelmsford. Grandfather was employed teaching young trainee engineers at the Marconi College, an institution of which he later became head and where he was regarded as a very good teacher. At first the family lived in Chelmsford itself, but later they moved into the neighbouring countryside, to a lovely sixteenth-century Essex longhouse called Water Hall near the straggling village of Little Baddow.

Little Baddow was the site of an anecdote about my grandfather which I think tells us something revealing about human nature. It was much later, during the Second World War, and Grandfather was out on his bicycle. A German bomber flew over and dropped a bomb (bomber crews on both sides occasionally did this in rural areas when, for some reason, they had failed to find their urban target and shrank from returning home with a bomb on board). Grandfather mistook where the bomb had fallen, and his first desperate thought was that it had hit Water Hall and killed his wife and daughter. Panic seems to have sparked an atavistic reversion to ancestral behaviour: he leapt off his bike, hurled it into the ditch, and *ran* all the way home. I think I can imagine doing that in extremis.

It was to Little Baddow that my Dawkins grandparents retired from Burma in 1934, to a large house called The Hoppet. My mother and her younger sister Diana first heard of the Dawkins boys from a girlfriend, breathless with Jane-Austen-style gossip about eligible young newcomers to the neighbourhood. 'Three brothers have come to live at The Hoppet! The third one is too young, the middle one is pretty good news, but the eldest one is completely mad. He spends all his time throwing hoops around in a marsh and then lying on his stomach and looking at them.'

This apparently eccentric behaviour of my father was in fact thoroughly rational – not the first or the last time a scientist's motives were uncomprehendingly called into to question. He was doing postgraduate research based in the Department of Botany at Oxford, on the statistical distribution of tussocks in marshes. His work required him to identify and count plants in sample quadrats of marshland, and throwing 'hoops' (quadrats) at random was the standard method of sampling. His botanical interest turned out to be among the things that drew my mother to him after they met.

John's love of botany had begun early, during one of the holidays from boarding school which he and Bill spent with their Smythies grandparents. In those days it was quite common for colonial parents to send their children, especially sons, to boarding school in Britain, and at the ages of seven and six respectively John and Bill were despatched to Chafyn Grove, a boarding school in Salisbury which I too was later to attend. Their parents would remain in Burma for another decade and more, and with no air travel would not see their sons even during most school holidays. So between terms the two little boys stayed elsewhere, sometimes at professional boarding homes for boys of colonial parents, sometimes with their Smythies grandparents in Dolton, Devon, where they often had their Smythies cousins for company.

Nowadays, such long-term separation of children from their parents is regarded with something approaching horror, but it was quite common at the time, accepted as an inevitable concomitant of empire, and indeed diplomatic service, when international travel was long, slow and expensive. Child psychologists might suspect that it did lasting damage. Both John and Bill, as it happened, ended up well-adjusted, very personable characters, but there may have been others less robustly equipped to come through such childhood deprivation. Their cousin Yorick, as I have already mentioned, was eccentric and possibly unhappy; but then, he went to Harrow, which

– to say nothing of the pressures of association with Wittgenstein – might explain everything.

During one of these school holidays with the grandparents, old Arthur Smythies offered a prize to whichever of his grandchildren could make the best collection of wildflowers. John won, and that boyhood collection became the nucleus of his own herbarium, setting him on the road to becoming a professional botanist. As I have said, a love of wildflowers was one of the things he later found in common with Jean, my mother. They also shared a love of remote and wild places, and a dislike of noisy company: they were not fond of parties, unlike John's brother Bill and Jean's sister Diana (who later married each other).

At the age of thirteen, John and then Bill left Chafyn Grove and were sent to Marlborough College in Wiltshire, one of England's better-known 'public' (i.e. private) schools, originally founded for the sons of clergymen. The regime was spartan; cruel, according to John Betjeman in his verse autobiography. John and Bill don't seem to have suffered in the way the poet did – indeed, they enjoyed it – but it may be revealing that, when Colyear's turn came some six years later, their parents decided to send him to a gentler school, Gresham's, in Norfolk. For all I know, Gresham's might have suited John better too, except that Marlborough had a legendary teacher of biology, A. G. ('Tubby') Lowndes, who probably inspired him. Lowndes has a number of famous pupils to his credit, including the great zoologists J. Z. Young and P. B. Medawar and at least seven Fellows of the Royal Society. Medawar was an exact contemporary of my father, and they went on to Oxford together, Medawar to read zoology at Magdalen and my father to read botany at Balliol. I have reproduced, in the web appendix, a historical vignette which is a transcript of a monologue by Lowndes, recorded verbatim by my father and almost certainly heard by Medawar in the same Marlborough classroom. I think it is of interest as a kind of anticipation of the central idea of the 'selfish gene', although it didn't influence me as I didn't discover it in my father's notebook until long after *The Selfish Gene* was published.

After his degree at Oxford, my father stayed on to do a post-graduate research degree – the one on tussocks that I mentioned earlier. He then decided on a career in the agriculture department of the Colonial Service. This necessitated further training in tropical agriculture at Cambridge (where his landlady had the memorable name of Mrs Sparrowhawk) and then – after becoming engaged to Jean – at the Imperial College of Tropical Agriculture (ICTA) in Trinidad. In 1939 he was posted to Nyasaland (now Malawi) as a junior Agricultural Officer.

CAMP FOLLOWERS
IN KENYA

JOHN'S posting to Africa hastened my parents' plans, and they were married on 27 September 1939 in Little Baddow church. John then left by ship for Cape Town, whence he travelled on to Nyasaland by train, and Jean followed in May 1940 in the flying boat *Cassiopeia*. Her rather dramatic journey took a week, with numerous landings for refuelling; one of them was in Rome, which caused some anxiety as Mussolini was teetering on the brink of entering the war on the German side, and had he done so at that point the *Cassiopeia*'s passengers would all have been interned for the duration of the war.

As soon as Jean arrived, John had to break it to her that he had been called up to join the King's African Rifles (KAR) in Kenya. The young couple had only a month of married life in Nyasaland (during which time, calculating backwards, I must have been conceived) before they had to leave. The Nyasaland Battalion was sending a convoy by road to Kenya, where they were to train. John somehow wangled permission to bypass the convoy and drive himself. What he did not have permission to do was take his bride with him. The colonial wives of Nyasaland were under strict orders to stay behind, or go to England or South Africa, when their husbands trekked north to the war. As far as she knows, my mother was the only one who disobeyed. My wonderful parents smuggled her into Kenya illegally – which caused problems later, as I shall tell.

On 6 July 1940, John and Jean, together with their servant Ali, who loyally accompanied them and was later to play a big part in my young life, drove off in 'Lucy Lockett', their old rattletrap Ford station wagon. They kept a joint diary of their journey, which I shall quote in what follows. They deliberately set off ahead of the convoy, in case they might break down and need rescuing: a prudent decision, given that the very first page of the diary mentions that the car had to be pushed by a gang of boys to get started at all. Day 4 of the journal records, after a successful bout of haggling for some gourds:

> This episode made us feel very cheerful, especially having won the battle and secured our gourds, and John was so hearty that he started the car before Ali was in the car & ripped off the door on a tree. This was very sad.

But even the mishap of losing a door didn't depress their young spirits, and the trio cheerfully made their way north, past ostriches and under giraffes, with Kilimanjaro on the horizon, sleeping by night in the back of the car, making a fire at each camp to scare the lions and cook delicious stews and pies in an improvised oven – the kind of ingenious invention my father delighted in throughout his life. From time to time they met up with the convoy. On one of these occasions the Commanding Officer, a

> big military gent . . . red hat and gold braid and minions, dived into an Indian shop, having commanded us to wait, and came out with a large bar of chocolate which he presented to me saying, 'A present for a little girl going on a big journey!' John ate the chocolate.

I wonder whether the chocolate was the genial commander's way of winking at the illegality of Jean's presence?

As they neared the Kenyan border,

> We were prepared to bury me under the bedding rolls, and have Ali sit on top when the Kenya frontier appeared. But the frontier never materialized, and after the most amazing and wonderful trip we found ourselves driving into Nairobi, and no one any the wiser. John deposited me in the Norfolk Hotel and drove off to join up – with Ali, who soon pinched an askari uniform and appointed himself a soldier.[1] Later he came out 'top' in an askari driver's course, thereby drawing attention to himself and causing John much embarrassment.

Despite this embarrassing triumph Ali never was officially a soldier, but he travelled around as my father's unofficial batman, accompanying him wherever he went, from training camp to training camp. At one of these, Nyeri, they coincided with the military funeral of Lord Baden-Powell, founder of the Boy Scouts. As a former scout himself, John was drafted in to be a pall-bearer and march beside the gun carriage. I have a photograph of him on this occasion (reproduced in the plate section) and I must say I think he looks very dashing in his KAR uniform, complete with khaki shorts, long socks, and the hat whose increasingly battered remains he was to wear for the rest of his life. Incidentally, the tall officer marching (out of step) next to him is Lord Errol of 'Happy Valley', soon to be murdered in the notorious and still officially unsolved 'White Mischief' case.

For Jean, the next three years were a time of more-or-less continuous migration as she camp-followed John's many postings,

---

[1] 'Askaris' was the name given to the African rank and file in the KAR.

in Uganda as well as Kenya. As she remarked in the private memoir that she wrote for the family much later,

> John was very clever at finding temporary homes for me near his different postings while he was training in the KAR. I did little jobs looking after people's children, and working in a couple of Prep schools, as well as being just a paying guest. Once John's commanding officer said when they had orders to go and take Addis Ababa that they'd better be quick or Jean Dawkins would be there first!

Among Jean's many kind hosts during this period were a Dr and Mrs McClean in Uganda, who took her in as a nursemaid for their toddler daughter 'Snippet'.

> The McCleans in Jinja were kind to me, and I trailed after Snippet doing this and that. The houses in Jinja were all around a golf course on the lake shore, and hippos used to play on the greens at night, belching and grunting, and marauding gardens too. There were droves of crocodiles, lazing in the water and basking at the shallow edges just below the falls, where I stupidly used to paddle. The crocs were funny keeping their jaws wide open so that their little pet bird friends could safely pick their teeth for them!

The symbiotic cleaner habit is now well described in coral reef fish. I wrote about it and the interesting evolutionary theory underpinning it in *The Selfish Gene*, but it hadn't occurred to me, until I read my mother's memoir much more recently, that there is a similar relationship between crocodiles and birds. I would expect the underlying evolutionary theory to be the same, best expressed in the mathematical language of game theory.

It was while staying with the McCleans that my mother got the first of her many bouts of malaria, which were to recur during her nine years in Africa and were one of the reasons for my parents' eventual decision to return to England. On one later occasion, when they were living in Nyasaland after the war, she has a vivid memory of hearing, through her fevered delirium, the urgent voice of Dr Glynn, senior physician of the Lilongwe hospital, saying: 'If they don't call John Dawkins quickly it may be too late.' Probably wrongly, she attributes her recovery to overhearing the doctor's fear that she was dying and her defiant resolve to prove him wrong.

However, one of her first suspected bouts of malaria at the McCleans' house turned out to have a different diagnosis:

> The doctor was a cheerful breezy chap and one day he said: 'You know what your trouble is, don't you?', and I said: 'malaria?', and he said, 'You're pregnant, my dear!' That was a shock, but we were delighted. Of course looking back it was very wrong of us in such an unpredictable and homeless situation. But then, had we been prudent and sensible and safe we would not have got Richard! So there! We took it in our stride, and I started making baby clothes and of course we were lucky. Luck stayed with us all the way. But now I realize it must have been hard for Richard later on being dragged all over the world, and may have been alarming to him. We made a list of how many times his little suit-case was packed in his first few years. Many nights were spent in the Kenya and Uganda Railway trains. Everywhere there were new faces and his early years must have been pitifully insecure.

I have found the list she made, covering my peregrinations during 1941 and 1942. She wrote it in a notebook, the 'blue book', now very tattered, in which she also recorded some of my childish sayings, and later those of my sister Sarah. The only place in the list that I remember is Grazebrook's Cottage, Mbagathi, near Nairobi, probably because we were there on two separate occasions. Here we

33

were the guests of Mrs Walter, her war-widowed daughter-in-law Ruby, and her little grandsons.

My mother's memoir continues:

Kenya, Uganda and Tanganyika are full of memories, many very happy and wonderful. But a lot of sorrow and fear and anxiety and loneliness after John went away for long periods and there was no news. Letters were very far between and tended to come in bunches with very old dates. I was often frightened and lonely and always anxious but we did have a lot of good kind friends and I was so lucky in that. Most notable were the Walters at Mbagathi who totally adopted Richard and me.

I was there when the telegram came to say that [Mrs Walter's son] John, who had just been home on leave, had been killed. Mrs Walter had been through it all before with her husband in the first world war when John was a baby. It was very very bad.

So we all concentrated on young William Walter and then later on posthumous Johnny. Richard had them as brothers and Mrs Walter as a granny for quite a while. She was a remarkable and splendid lady and she kept busy and positive. She concentrated on giving happy holidays to servicemen on leave and I used to be sent into Nairobi, ferrying in and out batches of soldiers and sailors and airmen, in Juliana who was not a very predictable form of transport. Juliana had two fuel tanks, she started on petrol, and then with luck switched over to paraffin. Once I only just survived the 20 odd miles home. An enormously fat huge naval cook, badly drunk I soon realized, who I'd fetched from the New Stanley Hotel fell asleep across the seat and leaned against me so heavily I could barely steer the car and I couldn't move him. It was very difficult.

I think those men really enjoyed the Walter ménage. They played with the children, did lots of man-about-the house little jobs for Mrs Walter who treated them as boys and fed wonderful meals. It was a real home for us all.

Richard and I built another mud hut at Mbagathi, a splendid

double one of two rondavels [the traditional circular form] elongated with a straight length between them. It was lovely.

These two huts with a shared roof only took about a week to construct. They constitute what I believe is my earliest memory.

Mrs Walter had by then purchased a bit of land. One day when she was clearing bushes with an African there was a huge explosion and the poor man had the back of one lower leg blown clean off by (we presumed) a first world-war left-over-mine. She was a very tall strong person and she lifted him into her box-body old banger and brought him home. We propped him and covered him, and she took him to Nairobi. He remained totally cheerful and chattered throughout. We could not believe such amazing bravery!

It is easy to forget that the First World War reached far down into sub-Saharan Africa. Tanganyika (plus Rwanda and Burundi) was in those days German East Africa and there was fighting in the area, including even naval battles in Lake Tanganyika between German boats on one side and those of Britain and Belgium on the other (the west coast of the lake was in the Belgian Congo). Elspeth Huxley, in her truly great novel *Red Strangers*, an epic saga of Kikuyu life, portrays the war through Kikuyu eyes as a mysterious and unspeakable aberration of the white men, in which Africans became horrifically caught up. Not only was it horrible, it was completely pointless, because the winning side didn't end up driving home any of the losers' cattle or goats.

Not all the shocks of this time were to do with wars, current or past.

Sometimes I was sent on Ruby's horse called Bonnie to take a message to the Lennox Browns' neighbouring farm. The first time

I went the house boy showed me into their big drawing-room while he called the Memsahib. The room was dark with chintzy curtains drawn against the bright sun and as I waited I suddenly realized that I was not alone. There was an enormous lioness stretched full-length on a sofa, who yawned at me! I was fairly paralysed. When Mrs Lennox Brown came in she smacked it and pushed it off the sofa. I gave my message and left.

My mother's picture of the incident, painted recently from her memory, appears in the picture section.

Later, Richard and William Walter used to play with two pet lion-cubs at another farm. They were about the size and heaviness of full-grown big Labradors (with short legs) and very rough and powerful. But he and William seemed to find it fun. We used to go for picnics up into the Ngong hills driving over the short mountain grass – no roads. Cool and high and splendid. But we were certainly stupid because there were buffaloes there in huge droves over those hills.

My next two memories are both of injections: the first by Dr Trim in Kenya and the second (more painful) by a scorpion, later in Nyasaland. Dr Trim was fortuitously well named, for he was presumably the one responsible for having me circumcised. Obviously I wasn't asked for my consent, but it seems my parents weren't either! My father, away at the war, knew nothing of it. My mother was simply informed as a matter of routine by a nurse that it was time for me to go for my circumcision, and that was that. Apparently it was the default presumption in Dr Trim's nursing home – as it may have been in many British hospitals of the time: at my various boarding schools, the numbers circumcised and uncircumcised were about even, and there was no obvious correlation with religion, or social position, or indeed anything else that I could detect. The situation is different in Britain today, and I understand that America is beginning to move in the same direction. A recent landmark case in a German court ruled that even religious circumcision of infants is a violation of the rights of those too young to give their consent. This German verdict will probably be overruled because of the shrieks of protest that to prevent parents circumcising their children is a violation of the parents' rights to practise their religion. Significantly, no mention of the child's rights. Religion enjoys astonishing privileges in our societies, privileges denied to almost any other special interest group one can think of – and certainly denied to individuals.

As for the scorpion, it gave me a painful rebuke for my deficiencies as a budding naturalist. I saw it crawling across the floor and I misidentified it as a lizard. How *could* I? Lizards and scorpions don't resemble each other in any respect that I can now see. I thought it would be fun to feel the 'lizard' run over my bare foot, so I stuck it in the animal's path. The next thing I knew was a burning pain. I screamed the house down and then I think I passed out. My mother tells me that three Africans heard my screams and came rushing in. When they saw what had happened, they took turns at trying to suck the poison out of my foot. That is a recognized emergency procedure for snakebite. I have no idea whether it is effective with scorpion stings but I am touched that they tried. I now actually have a horror of scorpions, such that I would not pick one up even if it had had its sting removed. As for the eurypterids, the giant marine scorpions of the paleozoic era, some of which reached six feet long . . .

I am often asked whether my African childhood prepared me to become a biologist, and the episode of the scorpion is not the only indication that the answer is no. Another story suggests the same, and I blush to tell it. Close to Mrs Walter's house when we were living there, a pride of lions had made a kill and some neighbours offered to take the whole household to watch them. We drove in a

safari car to within 10 yards of the kill where the lions were gnaw-
ing, or in some cases lying around as if they had already eaten too
much. The adults sitting in the vehicle were transfixed with excitement and wonder. But, my mother now tells me, William Walter and
I stayed on the floor, totally absorbed with our toy cars, which we
were driving around saying *vroom vroom*. We showed complete
indifference to the lions, despite the adults' repeated attempts to
arouse our interest.

What I lacked in zoological curiosity I seem to have made up for
in human sociability. My mother says that I was exceptionally
friendly, with no fear of strangers: an early talker with a love of
words. And despite my shortcomings as a naturalist I do seem to
have been an early sceptic. At Christmas 1942 a man called Sam
dressed up as Father Christmas and entertained a children's party in
Mrs Walter's house. He apparently fooled all the children, and finally
took his departure amid much jovial waving and ho-ho-ho-ing. As
soon as he had left, I looked up and breezily remarked, to general
consternation, 'Sam's gone!'

My father came through the war unscathed. I guess he was lucky
to be fighting not the Germans or Japanese but the Italians, who
perhaps had by then seen through their preposterously vainglorious
Duce and were sensible enough to have lost interest in winning. John
played his subaltern's part in armoured cars in the Abyssinian and
Somaliland campaigns and then, after the Italians were defeated, was
sent for training to Madagascar with the East African Armoured Car
Regiment, expecting to be posted to Burma. There he might have
met his younger brother Bill, who was by then a major in the Sierra
Leone Regiment, fighting the much more formidable Japanese and
later to be mentioned in despatches. However, in 1943 the government gave higher priority to John's agricultural than to his military
work, and he was recalled to civilian life, along with others of the
Nyasaland Agriculture Department.

The welcome news of his demobilization so excited Jean when she read it that she was nearly run over in the street, carrying me. She was fetching her mail, as usual, from the *poste restante* box in Nairobi. John's letter purported to be a description of a cricket match. But she had no interest in cricket, as John knew well, and he would never have bored her with it. It had to have a secret meaning. The couple had previously worked out a private code, and had used it several times before, because mail from army personnel in wartime was routinely opened and read by censors. Their code was a simple one: read only the first word of each line and ignore the rest. And the first words of the next three lines about the cricket match were '*bowler . . . hat . . . soon*'. Unfortunately the letter doesn't survive, but it is easy to imagine. 'Bowler' ostensibly referred to the cricket bowler, and John must have worked 'hat' in somehow (perhaps the umpire's Panama; my mother doesn't recall) and then 'soon' in some plausible comment about the match. What did it mean? Well, a bowler hat was the epitome of civilian dress – demob kit, civvy street. '*Bowler hat soon*' could only mean one thing, and Jean didn't need to be a crossword expert to discern it. John was about to be demobbed, and Jean nearly got herself and me run over in her excitement at the realization.

Actually getting back to Nyasaland, however, was not so easy. The illegality of Jean's original entry to Kenya now came back to haunt her. The dundridges[1] of the colonial government couldn't give her a visa to leave Kenya because, as far as their records showed, she

---

[1] My wife's and my private word for heartlessly rule-loving bureaucrats, a word that I am trying to introduce into the English language. It comes from a comic novel by Tom Sharpe, in which J. Dundridge epitomized the type. It's such a suitable-sounding word. For a new word to qualify for the *Oxford English Dictionary* it must be used sufficiently often in the written language, without definition or attribution. I speak from experience and am delighted to say that an earlier coining, 'meme', has met the criterion and is safely perched among the Ms. Please use dundridge and give it currency.

had never arrived. And Jean and John couldn't drive down together in the way that they had driven up, because this time John was under strict orders to travel with the army: he was not officially demobbed until he reached the Nyasaland Battalion's headquarters in its home country. So the couple had to leave Kenya separately, and Jean couldn't leave because she wasn't there. Mrs Walter was wheeled out to vouch for her existence and Dr Trim to vouch for mine – as, having brought me into the world, he was in a position to do. Finally, it was my legal birth certificate that did the trick, and the reluctant dundridges grudgingly stamped Jean's leaving papers. She and I, aged two, set off in a small plane of the kind that would today be called a puddle-jumper – pretty exciting puddles, no doubt, filled with crocodiles and hippos, flamingos and bathing elephants. We lost all our luggage when changing planes in Northern Rhodesia (now Zambia) but it soon didn't matter. My parents were delighted to find that their trunks, shipped by sea from England at the beginning of the war, had finally arrived in Nyasaland, having survived, presumably, a navy-escorted convoy, and containing, as my mother happily recalled in her memoir–

> All our half-remembered wedding-presents, and my new clothes. It was a tremendous home-coming, and Richard there to help explore the boxes.

THE LAND OF
THE LAKE

OUR life remained as peripatetic as it had been in Kenya. John and the other returnees from the army were used as stand-ins so that resident agricultural officers who had had no leave from their tropical duties since the beginning of the war could take a break in the balmy haven of South Africa. So John was posted to a different job, in a different part of Nyasaland, every few months. But, as my mother acknowledged, 'it was good fun and no doubt good experience for John, and we saw a lot of Nyasaland and lived in lots of interesting houses'.

Of this period, the house I remember best is the one at Makwapala, under Mount Mpupu near Lake Chilwa, where my father was in charge of an agricultural college and prison farm. The prisoners, who provided labour on the farm, seemed to have a good deal of freedom, and I remember watching them playing football with their toughened bare feet. My sister Sarah was born in Zomba hospital during this time, and my mother recalled that the Makwapala prisoners, some of them convicted murderers, 'used to queue up to be allowed to push her in her go-cart after tea'.

When we first arrived at Makwapala, we had to share the official Agricultural Officer's house with the outgoing family, whose return passage to England had been delayed a few weeks. They had two sons, the elder of whom, David, had the unpleasant habit of biting other children. My arms became covered with bite marks. On one occasion, at tea on the lawn, my father caught David at it and gently interposed his shoe to stop him. David's mother was outraged. She clasped the child to her bosom and roundly scolded my poor father. 'Do you have *no idea* of child psychology? Surely everybody knows that the very *worst* thing you can do to a biter is to stop him in mid-bite.'

Makwapala was a hot, humid, mosquito- and snake-infested place. It was too remote to enjoy a regular postal service, and the settlement had its own 'messenger', Saidi, whose daily job was to cycle the 15 miles to Zomba and back with the mail. One day Saidi didn't return; we learned that

> the unprecedented rain on Zomba mountain had roared down all the steep ravines washing great lumps of mountain and enormous rocks ahead of it. In Zomba Town, roads and bridges disappeared, and people in their cars, houses were marooned, and of course the road to Makwapala had washed away.

Saidi was safe, but I was sad that a nice man called Mr Ingram, who used to let me drive his car sitting on his lap, had been killed when a bridge that he was driving over washed away. 'Later', my mother wrote, 'we learned from local people that this sort of thing had happened before, though not in living memory. It was caused by some enormous snake-like creatures called Nyapolos, who got into the valleys and disrupted everything.'

I loved the rain. I think I perhaps picked up the sense of relief that people in a periodically dry country feel 'the day that the rains came down'. At the time of the great Nyapolos rain, having 'missed out on rain mostly', I was apparently, 'enchanted – he stripped off and rushed about in the downpour shouting with joy and going quite mad'. I still get a warm feeling of contentment in heavy rain, but I no longer like being out in it, perhaps because English rain is colder.

Makwapala is the site of my earliest coherent memories, and also of many of my parents' recordings of my sayings and activities. Here are just two of many:

> Come and look Mummy. I've found where the night goes to sleep when it's sun-times [darkness under the sofa].

> I measured Sally's bath with my ruler, and it said seven and ninepence, so she's very late for her bath.

Like all small children I was obsessed with pretending.

> No, I think I'll be an accelerator.

> Now you stop being the sea Mummy.

> I am an angel, and you're Mr Nye, Mummy. You say Good morning Angel. But angels don't talk, they just grunt. Now this angel's going to sleep. They always go to sleep with their heads under their toes.

I also enjoyed second-order meta-pretends:

> Mummy, let me be a little boy pretending to be Richard.

> Mummy, I'm an owl being a water wheel.

There was a water wheel near where we lived, which fascinated me. My three-year-old self tried to put together some instructions for how to make a water wheel:

> Tie a bit of string on the sticks all round, and have a ditch near and

very fast water in it. Now get a bit of wood and put a bit of tin on it for a handle and use it for the water to come. Then get some bricks for the water to go rushy down, and get a bit of wood and make it round and make a lot of things sticking out of it, then put it onto a long stick and that's a water wheel and it goes round in the water and makes a big BANG BANG BANG noise.

I suppose the following is zero-order pretending, for my mother and I both had to pretend to be ourselves:

Now you be Mummy and I'll be Richard and we're going to London in this garrimotor [most likely this Anglo-Indianism entered my family through my colonial grandparents and great-grandparents, but it may have spread from India throughout the Empire].

In February 1945, when I was nearly four, my parents recorded that I had 'never been known to draw anything recognizable'. This may have been a disappointment to my artistically gifted mother, who had been hired to illustrate a book when she was sixteen, and later attended art school. To this day I remain quite extraordinarily inept where visual art is concerned, and I have a blind spot even for appreciating it. Music is another matter entirely, as is poetry. I can easily be moved to tears by poetry and (slightly less easily) by music, for example the slow movement of the Schubert String Quintet, or some songs of Judy Collins and Joan Baez. My parents' notes show an early fascination with the rhythms of speech. They would listen in when I was having my afternoon rest at Makwapala.

> The wind blows in
> The wind blows in

> The rain comes in
> The cold comes in
> The rain comes
> Every day the rain comes
> Because of the trees
> The rain of the trees

Apparently I talked or sang to myself all the time, often in nonsensical but rhythmic cadences.

> The little black ship was blowing in the sea
> A little black ship was blowing in the wind
> Down down down to the sea
> Down in the meadows, a little black ship
> The little black ship was down in the meadows
> The meadows were down to the sea
> Down to the meadows, and down to the sea
> The little black ship down in the meadows
> Down in the meadows, down to the sea

I think that this kind of soliloquizing, experimenting with rhythms and permuting words perhaps only half understood, is common among small children. There is a very similar example in Bertrand Russell's autobiography, when he tells of eavesdropping on his two-year-old daughter Kate talking to herself, and hearing her say:

> The North wind blows over the North Pole.
> The daisies hit the grass.
> The wind blows the bluebells down.
> The North wind blows to the wind in the South.

My best guess is that my garbled allusion to Ezra Pound

in the following must have come from my parents' reading aloud.

> The Askari fell off the ostrich
> In the rain
> Huge sing Goddamn
> And what became of the ostrich?
> Huge sing Goddamn

My parents also record that I had a large repertoire of songs, which I would render, always correctly in tune, pretending to be a gramophone, sometimes with 'jokes' such as getting stuck in a groove and singing the same word over and over until the 'needle' (my finger) was pushed out of the groove. We had a portable, wind-up clockwork gramophone, of exactly the kind immortalized in Flanders and Swann's 'Song of Reproduction'.

> I had a little gramophone
> I'd wind it round and round.
> And with a sharpish needle,
> It made a cheerful sound.
>
> And then they amplified it
> It was much louder then.
> And used sharpened fibre needles,
> To make it soft again.

My father didn't buy fibre needles. Characteristically, he improvised with the thorns at the end of sisal leaves.

Some of my songs I think I got from records, some were gibberish made up by me on the spur of the moment like those quoted above, and some were from my parents. My father, especially, delighted in teaching me nonsense songs, often derived from his

own father, and many an evening rang to the strains of such gems as
'Mary had a William goat', 'Hi Ho Cathusalem, the harlot of
Jerusalem' or 'Hoky Poky Winky Fum', which I learned was sung
daily by my Smythies great-grandfather while lacing up his boots
and at no other time. I was once temporarily lost on a Lake Nyasa
beach, and was eventually discovered sitting between a pair of old
ladies in deckchairs regaling them with the Gordouli song, bawled
since 1896 by Balliol undergraduates as a mocking serenade over the
wall to the neighbouring college, Trinity, and a favourite of my
grandfather and father.

> Gordoooooooooli.
> He's got face like a ham.
> Bobby Johnson says so.
> And he ought to know.
> Bloody Trinity. Bloody Trinity.
> If I were a bloody Trinity man
> I would. I would.
> I'd go into the public rear,
> I would. I would.
> I'd pull the plug and disappear.
> I would. I would.
> Bloody Trinity. Bloody Trinity.

Well, it's scarcely great poetry and never normally sung sober,
but I suppose it is slightly intriguing to wonder what the old ladies
made of it. My mother reports that, despite being missionaries, they
seemed to be enjoying it. When I eventually got to Balliol myself in
1959, by the way, I discovered that the tune had changed for the
worse – having suffered a destructive memetic mutation and lost a
subtlety – at some point during the twenty-two years since my father
had left.

My gramophone metaphor was regularly pressed into service in a guileful attempt to postpone bedtime: the gramophone would run down, the song becoming slower and grinding down in pitch, and would need to be 'wound up'. This was indeed a part of everyday life, for we had no electricity and our clockwork gramophone had to be wound up at frequent intervals to play my father's collection of 78 rpm records: mostly Paul Robeson, whom I adore to this day, plus another great bass, Feodor Chaliapin, singing *Tom der Reimer* in German (I wish I could track down that recording, but iTunes has so far let me down) and some miscellaneous orchestral music including César Franck's *Symphonic Variations*, which I called the 'Dripping Water', presumably in reference to the piano part.

With no electricity, our houses were lit by paraffin pressure lamps. They had to be primed with methylated spirit to heat the mantle, then pumped up with paraffin vapour, whereupon they hissed comfortably through the evening. For most of our time in Nyasaland we didn't have a water closet either, and had to use an earth closet, sometimes in an outhouse. In other respects, however, we lived in great luxury. We always had a cook, a gardener and several other servants (known, I regret to say, as 'boys'), headed by Ali, who became my constant companion and friend. Tea was served on the lawn, with beautiful silver teapot and hot-water jug, and a milk jug under a dainty muslin cover weighted down with periwinkle shells sewn around the edges. And we had drop scones (Scotch pancakes) which, to this day, are my equivalent of Proust's *madeleine*.

We had bucket-and-spade holidays on the sandy beaches of Lake Nyasa, which is big enough to seem like the sea with no land visible on the horizon, staying in a nice hotel whose rooms were thatched beach huts. We also had a holiday in a borrowed cottage high up Zomba Mountain. One anecdote from this trip demonstrates my lack of critical faculty (and perhaps belies the story of my seeing through Sam's Father Christmas act when I was one). Playing hide and seek with a friendly African man, I searched one particular hut and he definitely wasn't there. Later I went back to the same hut and he was there, in a place where I had positively looked. He swore that he had been there all the time, but had made himself invisible. I accepted this explanation as more plausible than the now obvious alternative hypothesis that he was lying. I can't help wondering whether a diet of fairy stories filled with magic spells and miracles, including invisible men, is educationally harmful. But whenever I suggest such a thing today I get kicked around the room for seeking to interfere with the magic of childhood. I don't think I told my parents my Zomba Mountain hide-and-seek story, but I can't help feeling that I'd have been rather pleased if they had talked me through a version of Hume on miracles. Which do you think would be the greater miracle? The miracle that a man might tell a lie to amuse a gullible child? Or the miracle that he really did turn himself invisible? So, little Richard, now what do you think really happened in that hut, high on Zomba Mountain rearing up out of the plain?

Another illustration of childhood gullibility: someone had attempted to relieve my distress at the death of pets by telling me that animals, when they die, go to their own heaven called the Happy Hunting Ground. I believed this totally, and didn't even wonder whether it was also 'heaven' for the prey animals they hunted there. Once, in Mullion Cove, I met a dog and asked whose dog it was. I misheard the answer as 'Mrs Ladner's dog come back'. I knew that, before I was born, my grandmother had had a dog called Saffron, now long dead. I immediately presumed, with a credulous curiosity too mild to be even worth following up, that this dog was indeed Saffron, returned from the Happy Hunting Ground for a visit.

Why do adults foster the credulity of children? Is it really so obviously wrong, when a child believes in Father Christmas, to lead her in a gentle little game of questioning? How many chimneys would he have to reach, if he is to deliver presents to all the children in the world? How fast would his reindeer have to fly in order that he should finish the task by Christmas morning? Don't tell her point blank that there is no Father Christmas. Just encourage her in the unfaultable habit of sceptical questioning.

Christmas and birthday presents in wartime, thousands of miles from relatives and high streets, were inevitably limited, but my parents made up for it in ingenuity. My mother made me a magnificent teddy bear, as big as I was. And my father made me various ingenious contraptions including a lorry, which had under the bonnet (hood) a single real (and incongruously but delightfully not-to-scale) sparking plug. The lorry was my pride and joy when I was about four. My parents' notes show that I would pretend it 'broke down', whereupon I would:

Mend the puncture
Wipe the water off the stridibutor (distributor)
Fix the battery
Put water in the radiator
Tickle the carburettor
Pull the choke
Try the switch the other way
Fix the plugs
Put the spare batteries in properly
Put some oil in the engine
See if the steering is all right
Fill up with petrol
Let the engine get cool
Turn it over and have a look underneath
Test the pops by shorting the terminals [I now don't know what that meant]
Change a spring
Fix the brakes

Etc

Each item is followed through with appropriate motions and noises, and is followed by Ger er er er er Ger er er er er on the starter, which may, or usually may not, start the engine.

In 1946, the war having ended the previous year, we were able to go 'home' to England on leave (England was always 'home' even though I had never been there; I have met second-generation New Zealanders who follow the same nostalgic convention). We went by train to Cape Town where we were to board the *Empress* (I thought it was 'Emprist') *of Scotland*, bound for Liverpool. South African trains had an open walkway between carriages, with railings like a ship's that you could lean over to watch the world go by and catch the cinders from the horribly polluting steam engine. Unlike a ship's, however, these railings had to be telescopic so they could lengthen or shorten when the train went round a bend. Here was an accident waiting to happen, and indeed it did. I had hooked my left arm over the rail and didn't notice when the train started going round a bend. My arm was caught as the railings telescoped in, and there was nothing my stricken parents could do to free me until the long bend ended and the line straightened out again. At the next station, Mafeking, the train was halted while I was taken to hospital to have my arm stitched. I hope the other passengers were not annoyed by the delay. I have the scar still.

When we finally reached Cape Town, the *Empress of Scotland* turned out to be a dismal ship. It had been converted as a wartime troop-carrier: no cabins, but dungeon-like dormitories with three-high rows of bunks. There were dormitories for the men and separate dormitories for the women and children. There was so little space that they had to take turns doing things like getting dressed. In the women's dormitory, as my mother's diary records . . .

it was bedlam with so many small children. We dressed them and took them to the door and handed them to the relevant father waiting in a long queue to collect his own. And he took them off to queue for breakfast. Richard had regular trips to the ship's doctor for dressings to his arm, and of course half way through the three-week voyage I had a malaria bout and Sarah and I were put into the ship's hospital, and poor Richard was left alone in the dreadful dormitory. They wouldn't allow him to go with John or me, which was cruel.

I don't think we appreciated what a horrible time that whole journey must have been for Richard. And what a long effect it must have had. He must have felt that his whole world security had suddenly gone. And when we got to England he was quite a sad little boy, and had lost all his bounce. While we were looking out of the ship at Liverpool docks in the dark rain, waiting to go ashore he asked wonderingly 'Is that England?' and then quickly asked 'When are we going back?'

We went to my paternal grandparents at The Hoppet in Essex, which

in February was bitterly cold and spartan, and Richard's confidence ebbed and he took to having a stammer. He couldn't cope with his clothes. Having lived most of his life in very few garments, buttons and shoe-laces defeated him and the grand-parents thought he was backward: 'Can't he dress himself yet?' Neither we nor they having any child psychology books they set about getting some discipline going and he became quite a withdrawn little person and a bit paralysed. There was a ritual in the Hoppet that he must learn to say Good Morning when he came to breakfast and he was sent out of the room until it happened – His stammer got worse and none of us were happy. I am ashamed now that we allowed that grandparental behaviour.

Things were not much better with the maternal grandparents in Cornwall. I disliked almost all food, and would psych myself up to retch when grandparents made me eat it. Horrible, watery vegetable marrow was the worst, and I actually vomited into my plate. I think everyone was relieved when the time came for us to board the *Carnarvon Castle* at Southampton bound for Cape Town, and return to Nyasaland – not back to Makwapala in the south, but to the central district around Lilongwe. My father was posted first to the agricultural research station at Likuni, outside Lilongwe, and then to Lilongwe itself, now the capital of Malawi but then a small provincial town.

Both Likuni and Lilongwe are places of happy memory. I must have been interested in science by the age of six, because I can remember regaling my poor long-suffering little sister, in our shared bedroom at Likuni, with stories of Mars and Venus and the other planets, their distances from Earth and their respective likelihoods of harbouring life. I loved the stars in that most un-light-polluted place. Evening was a magically safe and secure time, which I associated with the Baring-Gould hymn:

> Now the day is over,
> Night is drawing nigh,
> Shadows of the evening
> Steal across the sky.
>
> Now the darkness gathers,
> Stars begin to peep;
> Birds, and beasts, and flowers
> Soon will be asleep.

I don't know how it came about that I knew any hymns at all, because we never went to church in Africa (although we did when staying with the grandparents in England). I suppose my parents must have taught me that hymn, along with 'There's a friend for littul chuldren, above the bright blue sky'.

Likuni was also where I first noticed, and was fascinated by, the long shadows of evening, which at the time had none of the foreboding evoked by T. S. Eliot's 'shadow at evening rising to meet you'. Today, whenever I hear Chopin's Nocturnes, I am transported back to Likuni and the secure, comforting feeling of evening when 'stars begin to peep'.

My father invented wonderful bedtime stories for Sarah and me, often featuring a 'Broncosaurus' which said 'Tiddly-widdly-widdly' in a high falsetto voice, and lived faaaaar away in a distant land called Gonwonkyland (I didn't finally take the allusion until undergraduate days when I learned about Gondwanaland, the great southern continent that broke up to form Africa, South America, Australia, New Zealand, Antarctica, India and Madagascar). We loved watching the luminous dial of his wristwatch in the darkness, and he would draw a watch on our wrists with his fountain pen, so we could keep track of the time under our mosquito nets during the comfortable night.

Lilongwe, too, was a place of precious childhood memory. The official house of the District Agricultural Officer was smothered in cascades of bougainvillea. The garden was filled with nasturtiums, and I loved to eat the leaves. Their unique, peppery taste, still encountered occasionally in salads, is the other candidate for my Proustian *madeleine*.

The identical house next door was the doctor's. Dr and Mrs Glynn had a son, David, of exactly my age, and we played together every day, in his house or mine or round about. There were dark blue-black grains in the sand, which must have been iron because we picked them up by dragging a magnet on a piece of string. On the verandah we made 'houses', with little rooms and corridors, by draping rugs and mats and blankets over upended chairs and tables.

We even equipped our verandah 'houses' with piped water, whose plumbing we made by sticking together hollow stems from a tree in the garden. Perhaps it was a *Cecropia*, but we called it a 'rhubarb tree', presumably deriving the name from a song that we liked to sing (to the tune of 'Little Brown Jug'):

> Ha ha ha. Hee hee hee.
> Elephant's nest in a rhubarb tree.

We collected butterflies, mostly yellow and black swallowtails, which I now realize were probably various species of the genus *Papilio*. David and I, however, didn't differentiate: we called them all 'Daddy Xmas', which he said was their proper name although it made no sense of their yellow and black colour scheme.

My butterfly habit was encouraged by my father, who made me a box for pinning them, using dried sisal instead of the cork favoured by professionals, and by my Dawkins grandfather – who was a collector himself – when he and my grandmother came to visit. They planned a grand tour of East Africa, calling on their sons in turn. They went first to Uganda to see Colyear, then made their way south to Nyasaland, through Tanganyika, as my mother recounted,

in a series of short-term local bus journeys, incredibly uncomfortably packed in with crowds of Africans and poor chickens with their legs tied, and enormous bundles of goods. But there was no transport further than Mbeya [in southern Tanganyika]. However, a young man with a little light aircraft offered to try to take them on. So they set off but got into bad weather and had to turn back. Meanwhile we had heard nothing from them at all. When their weather improved they tried again, flying low so that Tony [my grandfather, short for Clinton] could lean out and identify rivers and roads reading an old map as they went, and directing the pilot.

Grandfather would have been in his adventurous element. He loved maps. Also railway timetables, which he knew by heart and which came to constitute his only reading matter in extreme old age.

In Lilongwe everyone knew when a plane was coming about ten minutes before it arrived. This was because a local family kept pet crested cranes in their garden. These birds could hear an approaching plane long before people could and would start shrieking about it. Whether in fear or joy one didn't know! The regular weekly plane not being due, we wondered if it could be the grandparents when the cranes started shouting one day – so we went up to the air field, Richard and David on their tricycles, and we were in time to see the tiny plane arrive circling around the town twice before landing with enormous bumps and then Granny and Grandfather climbing out.

Nothing so obvious as Air Traffic Control, then. Just crested cranes.

It was in Lilongwe that we were struck by lightning. One evening a huge thunderstorm came. It was very dark and the children were having their suppers under their mosquito nets in the (wooden) beds. I was reading sitting on the floor and leaning against our so-called sofa (made of an old iron bedstead). Suddenly I felt as though a sledgehammer had landed on my head and I was completely flattened. It was a tremendous, carefully aimed blow. We saw that the wireless aerial and a curtain were on fire and we rushed into the children's bedroom to see if they were alright. They were totally unaffected and were chewing on their maize-cobs in a fairly bored sort of way!

History doesn't relate whether my parents extinguished the curtain fire before or after rushing into our bedroom to see if we were safe. My mother's memoir continues:

I had a long red burn all across my side where I had been leaning against the iron bed, and we discovered all sorts of other funny things later. Like a lump of concrete floor torn up and put onto the garage roof! The cook had a knife snatched out of his hand and was knocked over, a wire clothes-line was melted and the panes of glass in the sitting room were all splattered with molten wire from the radio aerial which totally disappeared, etc. etc. We now can't remember it all but it was dramatic.

My memory of that lightning strike is hazy, but I do wonder whether the cook's knife was really snatched out of his hand or whether he threw it in fright – as I would have. I do recall the multicoloured patterns made by some kind of residue all over the windows. And the actual moment of strike itself when the noise, instead of the usual boom boom de boom boom boom (which is mostly echoes) consisted of a single, prodigiously loud bang. There must have been a simultaneous very bright flash, but I have no memory of it.

Luckily it didn't leave us thunderstorm-shy because there were plenty of splendid ones in Africa. They were immensely beautiful, silhouetting mountain ranges black against brilliant-lit skies, all to the grand opera accompaniment of the sometimes almost non-stop thunder.

At Lilongwe we bought our first ever brand new car, a Willys Jeep station wagon called Creeping Jenny, to replace Betty Turner, the old Standard Twelve. I remember with nostalgic delight Creeping Jenny's exciting new-car smell. Our father explained to Sarah and me its advantages over all other cars, most memorable of which were the flat mudguards over the front wheels. He explained to us that these were especially designed to act as tables for us to put our picnic on.

At the age of five I was sent to Mrs Milne's school, a little one-room nursery school run by a neighbour. Mrs Milne couldn't really teach me anything because all the other children were learning to read, and my mother had already taught me to read; so Mrs Milne sent me off to one side with a 'grown-up' book to read to myself. It was too grown-up for me and, although I faithfully forced my eyes to travel over every word, I didn't understand most of them. I remember asking Mrs Milne what 'inquisitive' meant, but I couldn't muster enough of the stuff to keep asking her the meanings of words when she was busy teaching the other children. So I then

shared lessons with the doctor's son David Glynn taught by the doctor's wife. They were both bright, keen little boys and we think they probably learned a lot. Then he and David went on to the Eagle School together.

EAGLE IN THE MOUNTAINS

THE Eagle School was a brand new boarding school set high among the conifers of the Vumba Mountains, near the border with Mozambique, in Southern Rhodesia (now the sick joke dictatorship of Zimbabwe). I use the past tense because the school closed for ever during the conflicts that later beset that unhappy country. It was founded by Frank ('Tank') Cary, a former housemaster from the Dragon School in Oxford, I think the largest and arguably the best prep school in England, with a wonderful spirit of adventure and a remarkable list of distinguished alumni. Tank had come out to seek his fortune in Africa, and his school was a faithful scion of the Dragon. We had the same school motto (*Arduus ad solem*, a quotation from Virgil) and the same school song, to Sullivan's tune for 'Onward, Christian Soldiers': '*Arduus ad solem /* By strife up to the sun'. Tank had visited our family in Lilongwe when he was on a tour trying to drum up business from Nyasaland parents: mine liked him and decided that Eagle was the school for me, as did Dr and Mrs Glynn for David, and we went there together.

My memory of Eagle is hazy. I think I was there for only two terms, including the second term of the school's existence. I remember being there for the formal opening of the school, which was much talked about in advance as the 'Opening Day'. This mystified me because I took it to be an allusion to 'O God our help in ages past':

> Time like an ever-rolling stream,
> Bears all its sons away;
> They fly forgotten, as a dream
> Dies at the *opening day*.

Hymns made a big impression on me at Eagle, even 'Fight the good fight with all thy might', sung to a stupefyingly dreary tune more appropriate to dozing than fighting. All parents were told to equip their sons with a bible. My parents, for some reason, gave me *The Children's Bible*, which was not the same thing at all, and I felt rather left out and 'different'. In particular it was not divided up into chapters and verses, which I felt as a terrible deprivation. I was so intrigued by the biblical method of subdividing prose for easy reference that I went through some of my ordinary story books, writing in numbered 'verses' for them too. I have recently had occasion to look at the *Book of Mormon*, fabricated by a nineteenth-century charlatan called Smith, and it occurs to me that he must have had the same fascination with the King James Bible, laying his book out in verses and even imitating the sixteenth-century English style. Incidentally, it is a mystery to me why that last fact alone didn't instantly brand him a fake. Did his contemporaries think the Bible was originally written in the English of Tyndale and Cranmer? As Mark Twain cuttingly remarked, if you removed all occurrences of the phrase 'And it came to pass', the *Book of Mormon* would be reduced to a pamphlet.

My favourite book at Eagle was Hugh Lofting's *The Story of Doctor Dolittle*, which I discovered in the school library. It is now widely banned from libraries for its racism, and you can see why. Prince Bumpo of the Jolliginki tribe, steeped in fairy tales, desperately wanted to be the kind of prince that frogs magically turn into, or that falls in love with Cinderellas. Concerned that his black face might frighten any Sleeping Beauties he should chance to awaken with a princely kiss, he begged Doctor Dolittle to turn his face white. Well, it's easy enough to see now why this book, un-remarkable and uncontroversial in 1920 when it was published, fell foul of the shifting *Zeitgeist* of the late twentieth century. But if we must talk moral lessons, the splendidly imaginative Doctor Dolittle books, of which I think the best is *Doctor Dolittle's Post Office*, are redeemed of their touch of racism by their much more prominent anti-speciesism.

In addition to its school song and motto, Eagle took over the Dragon School's tradition of calling the teachers by their nicknames or Christian names. We all called the headmaster Tank, even when being punished by him. At the time I thought the name meant the sort of tank that holds water in your roof, but I now realize that it almost certainly referred to the relentlessly unstoppable military vehicle. Probably Mr Cary acquired, during his years at the Dragon, a reputation for dogged persistence, moving forward in a straight line regardless of obstacles. Other masters were Claude (also a migrant from the Dragon), Dick (who had the popular duty of handing out a blessed ration of chocolate during our afternoon rest every Wednesday) and Paul, a darkly jovial Hungarian who taught French. Mrs Watson, who taught the most junior boys, was 'Wattie' and the matron, Miss Copplestone, was 'Coppers'.

I cannot pretend that I was happy at Eagle, but I was probably as happy as a seven-year-old sent away from home for three months can expect to be. Most poignant was the fantasy which I think I indulged almost daily when Coppers used to do her quiet morning rounds of the dormitories and we were still dozing: I imagined that she would somehow magically be transformed into my mother. I prayed incessantly for this – Coppers had dark curly hair like my mother, so in my childish naivety I reasoned that it wouldn't have taken a very big miracle to effect the transformation. And I was sure the other boys would like my mother just as much as we all liked Coppers.

Coppers was motherly and kind. I like to think that her report on me at the end of my first term was not entirely lacking in affection: I had, she wrote, 'only three speeds: slow, very slow and stop'. She did scare me once, without the slightest intention of doing so. I had a horror of going blind, having once seen an African with white blank-staring eyes like the ends of hard-boiled eggs. I used to fret that one day I would become either totally blind or totally deaf and I decided, after much painful deliberation, that it was a close-run thing but that going blind was the worst thing that could possibly happen. The Eagle School was modern enough to have electric light, driven by our own generator. One evening, as Coppers was talking to us in the dormitory, the generator engine must have died. As the light faded into total darkness, I quavered fearfully, 'Have the lights gone out?' 'Oh no,' said Coppers with breezy sarcasm, 'you must have gone blind.' Poor Coppers, she little knew what she said.

I was also terrified of ghosts, which I pictured as fully articulated, rattling skeletons with gaping eye sockets, sprinting towards me down long corridors at immense speed and armed with pickaxes, whose blows they would aim with devastating precision at my big toe. I also had weird fantasies of being cooked and eaten. I have no idea where these awful imaginings came from. Not from any books I had read, and certainly not from anything my parents had ever told me. Maybe tall stories recounted by other boys in the dormitory – of the type that I was to meet at my next school.

For Eagle was also my first exposure to the boundless cruelty of children. I wasn't bullied myself, thank goodness, but there was a boy called Aunty Peggy who was mercilessly teased, seemingly for no better reason than his nickname. As if in a scene from *Lord of the Flies*, he would be surrounded by dozens of boys, dancing around him in a circle and chanting 'Aunty *Peggy*, Aunty *Peggy*, Aunty *Peggy*' to a monotonous playground tune. The poor boy himself was driven demented by this, and would blindly rush at his tormentors in the circle, fists flying. On one occasion we all stood around and watched him in a serious and prolonged fight, rolling around the ground, with a boy called Roger, of whom we were in awe because he was twelve. The sympathy of the crowd was with the bully, who was good-looking and good at games, not the victim. A shameful episode, all too common among schoolchildren. Eventually, and not before time, Tank put a stop to this mass bullying, with a solemn warning at the morning assembly.

Every night in the dormitory we had to kneel on our beds, facing the wall at the head, and take turns on successive evenings to say the goodnight prayer:

Lighten our darkness, we beseech thee, O Lord; and by thy great mercy defend us from all perils and dangers of this night. Amen.

None of us had ever seen it written down, and we didn't know what it meant. We copied it parrot fashion from each other on successive evenings, and consequently the words evolved towards garbled meaninglessness. Quite an interesting test case in meme theory, if you happen to be interested in such things – if you are not, and don't know what I'm talking about, skip to the next paragraph. If we had understood the words of that prayer, we would not have garbled them, because their meaning would have had a 'normalizing' effect, similar to the 'proofreading' of DNA. It is such normalization that makes it possible for memes to survive through enough 'generations' to fulfil the analogy with genes. But because many of the words of the prayer were unfamiliar to us, all we could do was imitate their sound, phonetically, and the result was a very high 'mutation rate' as they passed down the 'generations' of boy-to-boy imitation. I think it would be interesting to investigate this effect experimentally, but have not so far got around to it.

One of the masters, probably Tank or Dick, used to lead us in community singing, including 'The Camptown Races' and:

> I have sixpence, jolly jolly sixpence,
> Sixpence to last me all my life
> I've tuppence to lend and tuppence to spend
> And tuppence to take home to my wife.

In this next one we were taught to sound the 'r' in 'birds', for reasons that I didn't understand at the time, but perhaps it was presumed to be an American song:

> Here we sits like brrrds in the wilderness
> Brrrds in the wilderness
> Brrrds in the wilderness
> Here we sits like brrrds in the wilderness
> Down in Demerara.

Some of the Dragon School's famous spirit of adventure had been exported to Eagle. I remember one exciting day when the masters organized the whole school into a large-scale game of Matabeles and Mashonas (a local version of Cowboys and Indians, using the names of the two dominant Rhodesian tribes) which had us roaming through the woods and meadows of the Vumba ('the mountains of the mist' in the Shona language). Goodness knows how we managed not to get lost for ever. And although the school had no swimming pool (one was built later, after I left) we were taken to swim (naked) in a lovely pool at the foot of a waterfall, which was far more exciting. What boy needs a swimming pool when you have a waterfall?

I made one journey to Eagle by plane, quite an adventure for a seven-year-old travelling alone. I flew in a Dragon Rapide biplane from Lilongwe to Salisbury (now Harare), from where I was to go on to Umtali (now Mutari). The parents of another Eagle boy, who lived in Salisbury, were supposed to meet me and set me on my onward journey, but they failed to show up. I spent what seemed like a whole day (with hindsight it cannot have been that long) wandering around Salisbury airport by myself. People were nice to me, somebody bought me lunch, and they let me wander into hangars and look at the planes. Weirdly, my memory is that it was quite a happy day and I wasn't at all frightened of being alone or of what might happen to me. The people who were supposed to meet me finally turned up and I got to Umtali where, I think, Tank met me in his Willys Jeep station wagon, which I liked because it reminded me of Creeping Jenny and home. I've told this story as I remember it. David Glynn has a different memory, and I'm guessing there were two journeys, one with him and one on my own.

FAREWELL TO AFRICA

I N 1949, three years after their previous leave, my parents had another leave and we journeyed to England from Cape Town again, this time in a nice little ship called the *Umtali*, of which I don't remember much except for the lovely polished wood panelling and the light fittings, which I now think were probably art deco. The crew was too small to have a paid entertainments officer, so one of the passengers, a life-and-soul-of-the-party type called Mr Kimber, was elected to perform the role. Among other things, when we passed the Equator he organized a 'crossing the line' ceremony, in which Father Neptune appeared in costume complete with seaweed beard and trident. Mr Kimber also organized a fancy-dress dinner at which I was a pirate. I was jealous of another boy who came as a cowboy, but my parents explained that his admittedly superior costume was simply bought off the shelf, whereas mine was improvised and therefore really better. I understand the point now, but didn't then. One little boy came as Cupid, completely naked, with an arrow, and a bow which he threw at people. My mother came as one of the (male) Indian waiters, darkening her skin with potassium permanganate, which must have taken many days to wear off, and borrowing a waiter's uniform with its prominent sash and turban. The other waiters played along with the joke and none of the diners saw through her: not even me, not even the Captain when she deliberately brought him ice-cream instead of soup.

I learned to swim on my eighth birthday in the *Umtali*'s tiny swimming pool, made of canvas stretched between posts and erected on deck. I was so pleased with my new skill that I wanted to try it in the sea. So when the ship docked at Las Palmas in the Canary Islands in order to take on a large cargo of tomatoes and the passengers were put ashore for the day, we went to a beach where I proudly swam in the sea, with my mother vigilant on the shore. Suddenly she saw an abnormally huge wave about to break, as she thought, right over my diminutive, dog-paddling self. Gallantly she rushed, fully dressed, into the water to save me. In the event the wave lifted me harmlessly up – and then broke with full force on my mother, who was soaked from head to toe. The passengers were not allowed back on the *Umtali* until evening, so she spent the rest of the day in salt-wet clothes. Ungratefully, I have no memory of this act of maternal heroism, and the account I have given is hers.

The cargo of tomatoes must have been poorly loaded, because it shifted alarmingly at sea and the ship listed so far to starboard that our cabin porthole was permanently under water, causing my little sister Sarah to believe that we 'really *have* sunk now, Mummy'. Things became worse in the notorious Bay of Biscay, where the *Umtali* was seized by a spectacular gale, so strong it was hard to stand up. I excitedly rushed down to our cabin and pulled a sheet off my bunk to use as a 'sail' because I wanted to be blown along the deck like a yacht. My mother was furious, telling me – perhaps rightly – that I could have been blown overboard. Sarah's precious comfort blanket, 'the Bott', was indeed blown overboard, which would have been a serious tragedy but for our mother's prior fore-sight in cutting it in half so that she could keep a spare that had the right smell. I'm interested in the phenomenon of comfort blankets, though I never had one myself. They seem to be held in a position to be smelled while thumb- or finger-sucking. I suspect that there is a connection to the research of Harry Harlow on rhesus monkeys and cloth mother-substitutes.

We eventually docked in the Port of London and went to stay in a lovely old Tudor farmhouse called Cuckoos, opposite The Hoppet, which my paternal grandparents had bought to protect the land from developers. Living with us were my mother's sister Diana, her daughter Penny and her second husband, my father's brother Bill, on leave from Sierra Leone. Penny was born after her father, Bob Keddie, was killed in the war, as were both his gallant brothers – a terrible tragedy for old Mr and Mrs Keddie, who understandably then lavished their attention on little Penny, their only remaining descendant. They also were very generous to Sarah and me, her cousins, whom they treated as honorary grandchildren, and regularly gave us our most expensive Christmas presents and took us annually to a play or pantomime in London. They were rich – the family owned Keddie's Department Store in Southend – and possessed a grand house with a swimming pool and tennis court outside, and inside a lovely Broadwood baby grand piano and one of the first television sets. We children had never seen a television before, and we were enthralled to watch the blurry black-and-white image of Muffin the Mule on the tiny screen in the middle of the big, polished wood cabinet.

Those few months living as two families in one at Cuckoos provided the kind of magical memories that only childhood can. Beloved Uncle Bill made us giggle, calling us 'Treacle Trousers' (which Google now tells me is Australian slang for what the English call 'trousers at half mast') and singing his two songs, which we would frequently request.

Why has the cow got four legs? I must find out somehow.
I don't know and you don't know and neither does the cow.

And this one, to a sailor's hornpipe tune:

> Tiddlywinks old man, get a kettle if you can,
> If you can't get a kettle get a dirty old pan.

Penny's half-brother Thomas was born in Cuckoos while we were there. Thomas Dawkins is my double cousin, an unusual relationship. We share all four of our grandparents and hence all our ancestors except our immediate parents. Our proportion of shared genes is the same as for half-brothers, but we don't, as it happens, resemble each other. When Thomas was born the family hired a nurse, but she lasted only as long as it took her to see dear Uncle Bill in action making breakfast for the two families. He was on the stone-flagged kitchen floor, surrounded by a circle of plates into which he was throwing eggs and bacon in turn like dealing cards. This was before the days of 'health and safety' but it was more than the fastidious nurse could stand and she walked out, never to return.

Sarah, Penny and I went daily to St Anne's School in Chelmsford, the school that Jean and Diana had attended at the same age and under the same head teacher, Miss Martin. I don't remember much about it, except for the mincemeaty smell of school dinners, a boy called Giles who claimed that his father had lain down between the rails and let a train run over him, and the fact that the music master was called Mr Harp. Mr Harp had us singing 'Sweet Lass of Richmond Hill': 'I'd crowns resign to call her mine', but I didn't interpret it as 'I'd resign crowns to call her mine'. I heard 'crownsresign' as a single verb, which I guessed, from the context, must mean 'very much like'. I had made the same kind of misunderstanding with the hymn 'New every morning is the love / Our wakening and uprising prove'. I didn't know what 'our prisingprove' was, but evidently a prisingprove was an object anyone should be thankful to possess. The St Anne's school motto was quite admirable: 'I can, I ought, I must, I will' (not necessarily in that order, but it sounds about right). The adults in Cuckoos were reminded of Kipling's 'Song of the Commissariat Camels', and recited it with such a swing that I have never forgotten it:

> Can't! Don't! Shan't! Won't!
> Pass it along the line!

I was bullied at St Anne's by some big girls – not really badly bullied, but badly enough to provoke me to fantasize that, if I prayed hard enough, I could call down supernatural powers to give the bullies their come-uppance. I pictured a purplish black cloud with a scowling human face in profile, streaking across the sky above the playground to my rescue. All I had to do was *believe* it would happen; presumably the reason it didn't work was that I didn't pray hard enough – as when I prayed at the Eagle School for the metamorphosis of Miss Copplestone. Such is the naivety of the childhood view of prayer. Some adults, of course, never grow out of it and pray that God will save them a parking place or grant them victory in a tennis match.

I was expecting to return to Eagle after one term at St Anne's. While we were in England, however, my family's plans changed radically and I was never to see Eagle or Coppers or Tank again. Three years earlier, my father had received a telegram from England to say that he had inherited from a very distant cousin the Dawkins family property in Oxfordshire, including Over Norton House, Over Norton Park, and a number of cottages in the village of Over Norton. The estate had been much larger when it first came into the family, bought in 1726 by James Dawkins MP (1696–1766). He left it to his nephew, my great-great-great-great-grandfather Henry Dawkins MP (1728–1814), father of the Henry who eloped with the help of four hansom cabs galloping in different directions.

Thereafter it passed through many generations of Dawkinses, including the disastrous Colonel William Gregory Dawkins (1825–1914), a choleric Crimean War veteran who is said to have threatened tenants with eviction if they didn't vote his way, which was – oddly – liberal. Colonel William was irascible and litigious and squandered most of his inheritance suing senior army officers for insulting him: a drawn-out and futile process which benefited nobody except – as usual – the lawyers. Apparently a raving paranoid, he publicly insulted the Queen, assaulted his commanding officer Lord Rokeby in a London street, and sued the Commander-in-Chief, the Duke of Cambridge. Even more unfortunately, believing it to be haunted, he pulled down the beautiful Georgian Over Norton House and in 1874 built a Victorian replacement. His lawsuits drove him deeper and deeper into debt, forcing him to mortgage the Over Norton Estate to something more than the hilt, and he died in penury in a Brighton boarding house, living on the £2 per week allowed him by his creditors. The mortgage was eventually paid off by his unfortunate heirs in the early twentieth century, but only by dint of selling off most of the land, leaving the small nucleus that eventually passed to my father.

By 1945, the owner of what remained was Colonel William's great-nephew, Major Hereward Dawkins, who lived in London and seldom went near the place. Hereward, like William, was a bachelor, and he had no close relations bearing the name Dawkins. Evidently, when making his will, he looked up the family tree and lit upon my grandfather as the senior surviving Dawkins. His lawyer presumably advised him to skip a generation, and so he ended up naming my father, his much younger third cousin, as his heir. As things turned out, it was a brilliant choice, although he couldn't have known at the time that my father was ideally suited to preserve the land and make a go of it: the two of them had never met, and I don't think my father even knew of Hereward's existence when the telegram arrived in Africa, out of the blue.

In 1899 a long lease on Over Norton House had been given, as a wedding present, to a Mrs Daly. No doubt the rent vanished into the bottomless pit of Colonel William's debt repayments. Mrs Daly lived there in grand style with her family, a pillar of the local gentry and stalwart of the Heythrop Hunt, and my parents had no expectation that Hereward's legacy would change their lives. My father intended to rise through the ranks of the Nyasaland Department of Agriculture until he retired (or, as it would in fact have turned out, until the country became independent as Malawi).

When the *Umtali* docked in England in 1949, however, my parents received a piece of unexpected news: old Mrs Daly had died. Their immediate thought was that they should set about finding another tenant. But the possibility of leaving Africa and farming in England began to occur to them, and slowly gained favour in their minds. Jean's susceptibility to a dangerous strain of malaria was one reason, and I expect they were also attracted by the thought of English schools for Sarah and me. Their parents counselled against leaving Africa, as did the family lawyer. The Dawkins parents thought it was John's duty, in keeping with family tradition, to carry on serving the British Empire in Nyasaland, while Jean's mother was filled with dark forebodings that they would 'fail farming' as most people did. In the end, Jean and John went against all advice and decided to forsake Africa, live at Over Norton and take the estate in hand as a working farm – the first time after more than two centuries as parkland for the leisured gentry. John resigned from the Colonial Service, forfeiting his pension, and apprenticed himself to a series of English small farmers to learn the new skills he would need. He and my mother decided not to live in Over Norton House itself, but to divide it up into flats in the hope that it might pay for itself (lawyers' advice was to pull it down and cut their losses).

We ourselves would live in the cottage at the entrance to the drive, but it needed a lot of renovation, and while this was being done we did live – well, camp would be a better way of putting it – in a corner of Over Norton House.

I was still very keen on Doctor Dolittle, and my dominant fantasy during this brief interlude in Over Norton House was of learning, like him, to talk to non-human animals. But I would go one better than Doctor Dolittle. I would do it by telepathy. I wished and prayed and willed all the animals from miles around to converge on Over Norton Park, and me in particular, so that I could do good works for them. I did this kind of wishful praying so often, I must have been deeply influenced by preachers telling me that if you want something strongly enough you can make it happen; that all it takes is willpower, or the power of prayer. I even believed you could move mountains if your faith was strong enough. Some preacher must have said this in my hearing and, as is all too common with preachers, forgot to make the distinction between metaphor and reality clear to a gullible child. Actually, I sometimes wonder whether they even realize there is a distinction. Many of them don't seem to think it matters much.

My childhood games around the same time were imaginative in a science-fictiony way. My friend Jill Jackson and I played spaceships in Over Norton House. Each of our beds was a spaceship, and we hammed it up for each other for hour after happy hour. It is interesting how two children can cobble together a storyboard for a joint fantasy, without ever sitting down together to work out the plot. One child suddenly says: 'Look out, Captain, Troon rockets are attacking on the left flank!' and the other instantly takes evasive action before announcing his side of the fantasy.

My parents had by now formally withdrawn me from Eagle and set about finding a school for me in England. They would probably have liked to send me to the Dragon, which was close by in Oxford, so that I could continue with something like the 'adventurous' Eagle experience. But such was the demand for places at the Dragon that you had to have your name down at birth to get in. So instead, they sent me to Chafyn Grove in Salisbury (the English Salisbury, after which the Rhodesian one was named), where my father and both his brothers had been, and not a bad school in its own right.

Chafyn Grove and Eagle were both – I should explain to those unfamiliar with such British arcana – 'preparatory schools': 'prep schools', for short. What did they 'prepare' us for? The answer is the even more confusing 'public schools', so called because they are in fact not public but private – open only to those whose parents can pay their fees. Close to where I live in Oxford there is a school called Wychwood, which for some years had a delightful notice outside the gates:

Wychwood School for Girls (preparatory for boys).

Anyway, Chafyn Grove was the prep school to which I was sent from eight to thirteen, to prepare me for public school from thirteen to eighteen. I don't, by the way, think it occurred to my parents to send me to anything other than the kind of boarding schools that Dawkinses normally attended. Expensive, but worth making sacrifices for – that would have been their attitude.

UNDER SALISBURY'S SPIRE

BEGINNING at any new school is bewildering. On the very first day I became aware that there were new words to learn. 'Puce' puzzled me. I saw it written on a wall and wrongly thought it must be pronounced 'pucky'. I eventually worked out that it was derogatory, synonymous with 'wet', also a favourite word, both meaning feeble. 'Muscle' meant the opposite: 'I was born in muscle India, Africa is puce' (in that era, many children who went to that kind of school were born in one or the other of those areas coloured Imperial pink on the map of the world). 'Wig', in the same school dialect, meant penis. 'Are you a roundhead or a cavalier? You know, your *wig*, is it a mushroom or a bootlace?' Such anatomical details were not confidential anyway, for we had to line up naked every morning for a cold bath. As soon as the rising bell sounded, we had to leap out of bed, take off our pyjamas, pick up our towels and stumble to the bathroom, where one of the three baths was filled with cold water. We plunged in and out as quickly as we could, supervised by the headmaster, Mr Galloway. From time to time the same bell was used to rouse us in the middle of the night for a fire practice. On one such occasion I was so dizzy with sleep that I went mechanically into morning getting-up routine, took my pyjamas off and had reached the bottom of the fire escape completely naked and carrying my towel before I noticed my mistake – everyone else was wearing pyjamas, dressing gown and slippers. Fortunately it was summer. The cold baths were not the only baths we had, of course. We had a proper hot bath in the evening (I forget how many times per week) in which we stood up to be washed by a matron, which we quite liked, especially when it was the pretty under-matron.

It was a time of austerity, close enough to the end of the war for many things still to be rationed. The food, with hindsight, was pretty horrible. Sweets were among the goods rationed by the government, and this had the paradoxical effect – presumably to the detriment of our teeth – that we actually had more sweets than we otherwise would have, because our sweet ration was scrupulously handed out after tea. I gave most of mine away. Now that I think about it, why was the wartime sweet ration anything other than zero? Couldn't what little sugar survived the U-boats have been put to better use?

My feet were frequently cold, and I suffered terribly from chilblains. Smells are notorious triggers of memory, and the eucalyptus smell of the chilblain liniment with which my mother supplied me is irrevocably associated with Chafyn Grove and the torment of itchy toes. We were often cold in bed at night, and we tried to stave it off by putting our dressing gowns on our beds. There was a chamber pot under each bed to obviate the need to go along the corridor during the night. I wish I had known at the time the North Country word for this object: gazunder (because it goes under).

Only one master was still at Chafyn Grove from my father's time: H. M. Letchworth, a kindly old Mr Chips-like figure who had fought in the First World War and had once been joint headmaster. We called him Slush, but not to his face, because Chafyn Grove didn't have the Dragon/Eagle convention about nicknames. The only exception was during the annual Scout Camp, when he liked to be called Chippi, an older nickname which I think dated from long before when he had known Baden-Powell. He didn't like the name e Dawkins family have been members of the Chipping Norton set since the early ghteenth century, when my great-great-great-great-grandfather Henry Dawkins MP ilt a family mausoleum in St Mary's Church for, in the words of the inscription the memorial tablet (*below*), 'himself and his heirs'. Brompton's 1774 portrait of enry's family serves as backdrop to a family photograph taken in Over Norton house ound 1958. My Grandfather Dawkins, with his pink Leander tie, sits between his fe Enid and his daughter-in-law Diana. My sister Sarah is in front of him; uncle Bill behind him between Uncle Colyear and me. My father is on the far left. My mother between Enid and Colyear's wife Barbara.

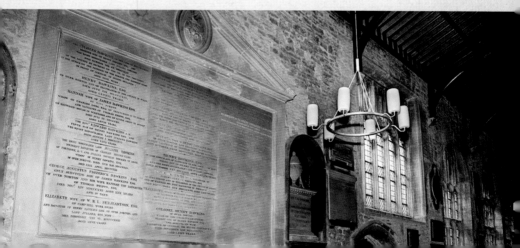

Is Zuleika Dobson among the spectators on the college barge as my grandfather Clinton G. E. Dawkins, leaning forward, prepares to row for Balliol?

My grandfather's education as an undergraduate (*right*) was supported by his uncle (later Sir) Clinton Edward Dawkins (*above*), whose freethinking views were celebrated in the Balliol Rhymes.

My father (*above*) and his rugby-playing brother Bill (*right*) followed their father and several other Dawkinses to Balliol after an idyllic childhood in the forests of Burma.

Above: The Smythies family at Dolton, Devon. *Top*: my paternal grandmother Enid, with dog and book, sits by her mother (in the very fine hat), brother Evelyn (with tennis racquet) and father (in panama hat), along with two unidentified guests. *Below*, Smythies cousins around 1923. Sitting on the ground, from right to left, are Bill, Yorick, John and Yorick's sister Belinda. Colyear is in his mother's arms.

Opposite: Evelyn Smythies' wife Olive was known as 'Tiger Lady' from her disagreeable hobby of shooting tigers. Her son, my father's first cousin Bertram Smythies, took a less destructive and more literary interest in the natural world.

THE AUTHOR ON AN ELEPHANT

RHINOCEROS HORNBILL

Frontispiece

THE BIRDS OF BORNEO

BY

BERTRAM E. SMYTHIES

B.A., M.B.O.U.

Overseas Forest Service, Sarawak

with special chapters by TOM HARRISSON, D.S.O., O.B.E.,
Curator of the Sarawak Museum, LORD MEDWAY, formerly
Technical Assistant, Sarawak Museum, and J. D. FREEMAN,
Ph.D., Reader in Social Anthropology at the National
University, Canberra, Australia

With 50 plates in colour by
COMMANDER A. M. HUGHES
O.B.E., R.N. (retd.)

and

93 photographic plates (2 in colour) by
LORD MAN THO, A.R.P.S., BRIAN BORRIDGE, A.R.P.S.,
E. O. H. ALLEN, R. D. MOLESWORTH,
R. E. W. IRVING AND HUGH GIBB

OLIVER AND BOYD

EDINBURGH : TWEEDDALE COURT
LONDON : 39A WELBECK STREET, W.1

1960

My maternal grandfather 'Bill' Ladner (seated third from left in the picture opposite), was among a group of naval officers sent to Ceylon to help build a wireless station during the First World War. Was the dog the station mascot? It seems to be the same dog my grandmother Connie is petting. The family returned to England when my mother Jean (*left*) was three. They lived in Essex (*above right*: my mother has her arms around a little friend) and spent their holidays at Mullion in Cornwall: here on the beach my Aunt Diana is holding hands with her mother and sister.

Above, my grandfather Ladner, a wireless engineer employed by Marconi and the author of the standard textbook on short wave wireless communication, demonstrates some equipment to visiting Arab royalty. He first met my grandmother in Cornwall while working at the Poldhu Wireless Station. Some of the thick slates used by the station as insulated instrument boards ended up as paving stones at our family house at the neighbouring Mullion Cove.

Slush. One Latin lesson the word *tabes* appeared in the vocabulary that we had to learn. Mr Letchworth was testing us and when the time came for a boy to translate *tabes* ('slush' in the context of the text we were reading) we all started sniggering. Mr Letchworth told us sadly that the name stemmed from that very passage of Livy ('All those years ago . . . that very sentence . . . all those years ago . . .'), though he never told us how it had come to stick to him.

The headmaster, Malcolm Galloway, was a formidable figure (maybe headmasters become formidable *ex officio*) whom we called Gallows. As befitted his nickname, he was not reluctant to use the extreme penalty, which in the case of Chafyn Grove was the cane. Unlike Eagle's 'bacon slice' beatings with a ruler, Gallows with the cane really hurt. He was reputed to have two canes, Slim Jim and Big Ben, and the punishment varied between three and six strokes depending on the severity of the misdemeanour. I never had Big Ben, thank goodness, but three strokes with Slim Jim was painful enough and caused bruises which we used to show off with pride, like battle scars, in the dormitory afterwards. They took several weeks to fade, turning from purple to blue to yellow on the way. Boys joked about stuffing an exercise book down the pants to soften the blows, but of course Gallows would have detected that instantly and I am sure it was never really tried.

Nowadays corporal punishment is illegal in England, and hind-sight suspects teachers who employed it of cruelty or sadism. I am convinced that Gallows was guilty of neither. We have here an example of the speed with which customs and values change – an aspect of what I called, in *The God Delusion*, the 'shifting moral *Zeitgeist*'. Not under that name, the shifting moral *Zeitgeist* over a great span of history is massively documented by Steven Pinker in *The Better Angels of our Nature*.[1]

---

[1] Steven Pinker, *The Better Angels of our Nature: Why Violence has Declined* (New York, Viking, 2011).

Gallows was capable of great kindness. He would go around the dormitories before lights-out like a genial uncle, cheering us up, calling us by our Christian names (then only: never during the school day). One evening Gallows noticed the *Jeeves Omnibus* on a shelf in my dormitory, and he asked whether any of us knew P. G. Wodehouse. None of us did, so he sat down on one of the beds and read a story to us. It was 'The Great Sermon Handicap', and I suppose he must have spread it over several evenings. We *loved* it. It has remained one of my favourite Jeeves stories, and P. G. Wodehouse one of my favourite authors, read, reread, and even parodied to my own purpose.

Every Sunday evening, Mrs Galloway used to read to us in the family's private sitting room. We had to leave our shoes outside and we all sat on the floor, cross-legged amid a faint smell of damp socks. She would read a chapter or two each week, and would get through a book a term. They were usually stirring adventure stories like *Moonfleet* or *Maddon's Rock* or *The Cruel Sea* (the 'cadet edition' with the sex scenes removed). One Sunday, Mrs Galloway was away and Gallows read instead. We had reached the bit of *King Solomon's Mines* where the gallant pith-helmeted heroes were confronted with the twin mountains called Sheba's Breasts (fascinatingly, this name was censored in the film version starring Stewart Granger, a version which, bizarrely, included a woman on the expedition). Gallows paused to explain to us that these mountains were the Ngong Hills. (*I say, you fellows, that's utter rot. Gallows is just showing off that he's been to Kenya. King Solomon's Mines wasn't set in Kenya at all. Race you upstairs to the dorm.*)

When there was a severe thunderstorm at night, Gallows went up to the most junior dormitory, switched the light on and comforted the tinies (small enough to be allowed teddy bears) who might have been frightened. Halfway through each term, on 'Going Out Sunday' when parents came to take their sons out for the day, there were always one or two boys with no visitors, perhaps because their parents were abroad or ill. It happened to me once. Mr and Mrs Galloway took us out with their own children, in their big old 1930s touring car called Grey Goose and their little Morris 8 called James. We had a lovely picnic by a weir, and it makes me almost tearful remembering how kind they were to us, especially given that they might have preferred a day out with just their own children.

But as a teacher, Gallows was frightening. He would bellow at the top of his powerful voice and his stentorian scorn could be clearly heard in all the other classrooms throughout the school, provoking conspiratorial smiles between us boys and the other masters. 'What do you do when you meet "*ut* with the subjunctive"? . . . STOP AND THINK!' (Though, come to think of it, rules like that are not the way language really works.) Mr Mills, one of several who taught Latin, was more frightening still: too alarming even to have a nickname. He had a menacing presence and insisted on total accuracy and flawless handwriting: one mistake and we had to write the whole passage out again. Miss Mills – no relation – plump, sweet and motherly, with pigtails tied up in a sort of halo around the back of her head, taught the tinies and called us all 'dear'. Mr Dowson, the jovial, bespectacled maths master, was nicknamed Ernie Dow. We none of us knew where the 'Ernie' came from until one day he read us a poem and ended up telling us the author: Ernest Dowson, of course. I don't remember which poem it was, possibly 'They are not long, the weeping and the laughter', but it was certainly wasted on us anyway. Ernie Dow was a good teacher who, in his faintly northern accent, taught me most of the calculus I was ever to know. Mr Shaw didn't have a nickname, but his teenage daughter was called 'Pretty Shaw' for no better reason than to justify the puerile joke that inevitably followed when anybody said 'I'm pretty sure . . .' There was a continuous turnover of young masters, presumably students waiting to go up to university or just come down from it, whom we mostly liked, probably just because they were young. One of these was a Mr Howard, Anthony Howard, who later became a distinguished journalist and editor of the *New Statesman*.

My first term, in Form II, I was taught by Miss Long, a thin, angular lady of middle age with straight hair and rimless glasses, very kind like most of the teachers. Apart from Form II, she mainly taught the piano. Indeed, my first music lessons were with her, and I boasted to my parents that my progress was much faster than it was. Since the truth was sure to emerge eventually, what was the point of such boasting? I shall never know.

It became apparent that, if my parents had been pessimistic about the academic standards of Eagle in Southern Rhodesia, they were wrong. I had been only average among my contemporaries at Eagle, but found myself well ahead when I first arrived at Chafyn Grove. Embarrassingly so, to the extent that, since academic ability was not admired, I would pretend to know less than I did. Asked for the meaning of a Latin or French word, for example, I would pretend to um and er uncertainly, rather than instantly show what I knew and risk losing face with my peers. This tendency became positively illogical the following year in Form III, when I foolishly decided that, since the muscle boys who were good at games were mostly not very good academically, the only way for me to become good at games was to perform poorly in class. Actually, now that I think about it, that attitude was so stupid it's pretty self-evident that I didn't deserve to do well in class anyway.

I was evidently very confused about what it meant to be good at games. There were three brothers, Sampson *ma*, Sampson *mi* and Sampson *min* (*major*, *minor* and *minimus*), who were all good, especially Sampson *min* who was brilliant at all sports and once 'carried his bat' all the way from opening to when he ran out of partners, and then made a miraculous catch at silly-mid-on. It ludicrously occurred to me that the resemblance of the Sampson name to that of the famous muscleman of the Bible could be no accident. My naive mind conjectured that the Sampsons must surely have inherited their sporting prowess, if not from the biblical hero himself, then from some medieval strongman ancestor who had earned the name in the same kind of way as 'Smith' or 'Miller' – or, indeed, 'Armstrong', which really does derive from a nickname for a man with strong arms. Among many other things I got wrong here was my assumption that noticeable hereditary qualities go back more than a couple of generations – the same 'Tess of the D'Urbervilles' error I mentioned in the first chapter.

The Sampsons' father, who had only one eye – the other one having been pecked out by a heron (or so we were implausibly told) – owned a farm in Hampshire, on which the Chafyn Grove Scout Troop held its annual camp, supervised by Slush, with assistance from Gallows and a portly gentleman called Dumbo who was drafted in for the occasion. Scout camp was a high spot of the year for me. We put up our tents, dug our latrines, and made a hearth for our fire on which we cooked delicious dampers and twists (lumps of dough charred in the fire). We learned how to lash sticks together with elegantly turned sisal string, and lashed up all manner of useful camp furniture, from mug trees to clothes horses. We sang songs around the camp fire, special Scout songs like 'Dai's got a head like a ping pong ball', taught us by Slush/Chippi and not difficult to learn, most of them being very short:

> Gaily sings the donkey, as he goes to grass.
> Who knows why he does so, because he is an ass.
> Ee aw. Ee aw. Ee aw, ee aw, ee aw.

Some of them had no tune and were yells of solidarity rather than songs:

There ain't no flies on us.
There ain't no flies on us.
There may be flies
On some of you guys
But there ain't no flies on us!

Pièce de résistance was an epic saga about a bad egg, sung by Chippi. I've reproduced it in the web appendix, in the sentimental hope that some of my readers might want to sing this otherwise forgotten song round a camp fire, and metaphorically stir the ashes of Henry Murray Letchworth MA (Oxon), Royal Dublin Fusiliers, alias Slush, alias Chippi, the benign and wistfully sad Goodbye-Mr-Chips patriarch of Chafyn Grove. In 2005, for my father's ninetieth birthday party in the Master's Lodgings, Balliol College, I faithfully wrote out the egg song for a bravura perform- ance by the lovely soprano Ann Mackay and her piano accompanist, and my father jovially, if less tunefully, joined in.

At Scout camp we earned badges for accomplishments such as axemanship, knot-tying, semaphore and Morse code. I was good at Morse, using a technique perfected by my father in wartime Somaliland when signalling from his armoured car. For each letter, you learn a phrase beginning with that letter. Single- syllable words represent dots, longer words dashes. G, for instance, was 'Gordon Highlanders go' – dash dash dot. I could construct no such mnemonics for semaphore and that was perhaps why I was bad at it. Or it may have been because I have low spatial intelligence: I do well on IQ tests until I hit the spatial rotation questions at the end, and they pull my score right down.

The other high spot of the year was the annual school play, always an operetta, always produced by Slush, in a tradition that had been going at least since my father's time. My uncle Bill later explained to me that he was 'auditioned for the part of a bulb, but was found wanting'. The lead roles went to boys who could sing, and I was one. *The Willow Pattern Plate*, in which I played the female lead in my last year, was typical. The scenery backdrop consisted of a large painting of the famous blue plate design. The pagoda was the residence of the royal princess; she had died, and to avert the threat of a republic, the three little men on the bridge had long conspired to keep her death a secret. Their plot was threatened when a handsome Tartar prince sent word that he was galloping on his way as a suitor. At this point I, as the village maiden, appeared and sang my big number describing, with hammed-up histrionic gestures towards the scenery, the blue ceramic world in which we all lived:

> Blue is the sky above my aching head.
> The grass is blue beneath my weary feet.
> Blue are the trees that o'er the blue path shed
> A deeper shade of everlasting blue.
> And all the world is clothed in robes of blue.
> The restless sea is of the self-same hue.

That last line is quite witty (wasted on us schoolboys, of course), and I'd like to think it got a laugh from the adult audience, which consisted almost entirely of devotedly tolerant parents plus the reporter from the *Salisbury Chronicle* (who, by the way, gave me a very nice but undeserved notice).

> The royal pagoda glistens in the sun.
> The footballs grow on yon preposterous tree.
> (*The song has several verses more to run*
> *But that's the lot in my poor memory.*)

The three little men on the bridge seized their opportunity and bundled me into the pagoda to impersonate the dead princess, just in time before the Tartar prince bounded onto the stage with black moustache drawn on his face and sword drawn from its scabbard. I can't remember how the happy ending was achieved, but the prince ended up slinging me over his shoulder in a fireman's lift and carrying me off back to Tartary.

Moments of acute embarrassment linger in memory and wring an audible groan out of me when I recall them. At Chafyn Grove we had a sit-down tea every day, where we ate bread and butter. While we were lining up to file into the dining room the duty master would sometimes read out a list of names, supplied by a boy whose birthday it was. The named invitees would drop out of the line and go to a special table set aside for birthdays at the end of the dining hall, laden with birthday cake, jellies and other good things sent by the loving mother. I understood the principle, and I understood about supplying the duty master with the list of your friends' names. That was very clear. What slipped my attention was the small point that you had to arrange for your mother in advance to send the cake and jelly. On my birthday – perhaps my ninth – I wrote out the list of my friends and gave it to the duty master, who read it aloud. My chosen friends walked eagerly into the dining room, surveyed the empty table and . . . even after all these years embarrassment prevents me from describing the scene any further. What still baffles me is why it never occurred to me to wonder where the cake was supposed to come from. Perhaps I vaguely thought the school cook would provide it. But even so, shouldn't I have wondered how the cook was supposed to know it was my birthday? Perhaps I thought it materialized by supernatural magic, like sixpenny bits when you put a tooth under your pillow. As with my Zomba Mountain hide-and-seek story, this incident reveals a sad lack of anything remotely approaching critical or sceptical thinking in my childhood years. Even though I find these examples embarrassing, the lack of ability to think through the *plausibility* of things is a human trait common enough to be interesting. I'll return to the theme.

I was an exceptionally untidy and disorganized little boy in my early years at Chafyn Grove. My first school reports dwelt insistently on the theme of ink.

Headmaster's Report: He has produced some good work and well deserves his prize. A very inky little boy at present, which is apt to spoil his work.

Mathematics: He works very well but I am not always able to read his work. He must learn that ink is for writing, not washing, purposes.

Latin: He has made steady progress but unfortunately when using ink his written work becomes very untidy.

Miss Benson, my elderly French teacher, somehow managed to omit the ink *Leitmotiv*, but even her report had a sting in the tail.

French: Plenty of ability – a good pronunciation and a wonderful facility in escaping work.

Ink? Well, what do you expect if you equip every desk with an open inkwell and give the children dip pens that might have been designed to flick ink all over the room, or at least to deposit great round glistening drops of ink all over a page – drops which I would then draw out in spider shapes, or turn into Rorschach blots by folding the paper? No wonder the row of washbasins was strewn with pumice stones (we thought they were pummy stones) for cleaning ink stains off fingers. I'm afraid the ubiquitous ink somehow managed to spread itself over more than just my exercise books. I would desecrate printed textbooks too. I'm not talking about changing Kennedy's *Shorter Latin Primer* to *Shortbread Eating*

Primer: everyone did that automatically, of course. My ink habit went further than that. I doodled all over the school textbooks, filling in letters with ink or drawing little cartoon figures in the top right-hand corner of the pages so that they moved cinematically when you flicked through the book. The books didn't belong to us: we had to hand them back at the end of every term ready for the next cohort to inherit. And I knew I would be in trouble when the time came for me to hand in my ink-encrusted textbooks. Worrying about this kept me awake at nights, it made me seriously unhappy and put me off my (admittedly pretty nasty) food, yet I still went on doing it. I recognize that there is a sense in which the book-desecrating child was the same individual as my present bibliophile self, but this perverse childhood behaviour is beyond my understanding today. As is my erstwhile reaction, and that, I suspect, of just about all my contemporaries now, to bullying.

Much of the apparent bullying was pure *braggadocio*, futile threats whose emptiness was attested by their invocation of an indefinite future: 'Right! That does it. I'm putting you on my *beating-up list*' was about as nebulous a threat as 'You'll go to hell when you die' (though, alas, not everybody treats the latter threat as nebulous). But there was real bullying too, the especially unkind form of bullying where gangs of sycophantic henchmen rally around a bullying leader, courting his approval.

The 'Aunty Peggy' of Chafyn Grove was even more seriously bullied than the one at Eagle. He was a precociously brilliant scholar, large, clumsy and ungainly, with an unharmonious, prematurely breaking voice and few friends. I won't mention his name in case he should happen to read this and the memory is still painful. He was an unfortunate misfit, an ugly duckling doubtless destined for swanhood, who should have aroused compassion, and would have done in any decent environment – but not in the Goldingesque jungle of the playground. There was even a gang bearing his name, the 'anti- —— gang', the sole purpose of which was to make his life a misery. Yet his only crime was to be awkward and gangling, too uncoordinated to catch a ball, unable to run except with a graceless staggering gait – and very, very clever.

He was a day boy, which meant that he could escape to his home each evening – unlike today's victims of bullying, who are pursued beyond the school gates on Facebook and Twitter. But there came a term when, for some reason – perhaps his parents went abroad – he became a boarder. And then the fun really took off. His torment was exacerbated by the fact that he couldn't stand the cold baths. Whether it was the cold water or the nakedness I don't know, but what the rest of us took in our stride reduced him to a state of whimpering, abject horror, clutching his towel to him, trembling uncontrollably and refusing to let go. It was his Room 101. Finally, Gallows took pity on him and excused him the cold baths. Which, of course, did wonders for his already rock-bottom popularity among his peers.

I cannot even begin to imagine how human beings could be so cruel, but to a greater or lesser extent we were, if only through failing to stop it. How could we be so devoid of empathy? There's a scene in Aldous Huxley's *Eyeless in Gaza* where men recall with shame and bewilderment their bullying of a similar ugly duckling in the dormitory at their old school. Perhaps the guilt, which I and presumably all of my Chafyn Grove friends who remember the episode feel, goes some little way towards helping us understand how guards in concentration camps could possibly have done what they did. Could the Gestapo represent a sort of retention into adulthood of a psychology that is normal in children, giving rise to an adult psychopathy? That's probably too simplistic, but my adult self remains baffled. It's not as though I was without empathy. Doctor Dolittle had taught me to empathize with non-human animals to a degree that most people would consider excessive. At the age of about nine I was fishing from a boat out of Mullion harbour with my grandmother when I had the misfortune to catch a mackerel. I was immediately filled with tearful remorse and wanted to throw it back. I cried when I wasn't allowed to. My grandmother was kind and consoled me, but not kind enough to let me throw the poor creature back.

I also empathized, arguably again to an excessive extent, with schoolfellows who were in trouble with the authorities. I would go to ridiculous lengths – actually rather courageously foolhardy lengths – to try to exculpate them, and I have to regard this as evidence of empathy. Yet I didn't lift a finger to stop the grotesque bullying that I have just described. I think this was partly due to a desire to remain popular with dominant and popular individuals. It is a hallmark of the successful bully to have a posse of loyal lieutenants, and again we see this brutally manifested in the verbal cruelty and bullying that has become epidemic on internet forums, where the abusers have the additional protection of anonymity. But I don't recall feeling even secret pity for the victim of the bullying at Chafyn Grove. How is that possible? These contradictions trouble me to this day, together with a strong feeling of retrospective guilt.

Once again, as with the ink, I am struggling to reconcile the child with the adult that he became; and the same struggle, I suspect, arises with most people. The apparent contradiction arises because we buy into the idea that the child was the same 'person' as the adult: that 'the child is father of the man'. It is natural to do this because of the continuity of memory from day to day and, by extension, from decade to decade even though, we are told, no physical molecule of the child's body survives the decades. Since I kept no diaries, it is precisely that continuity that makes it possible for me to write this book. But some of our deepest-thinking philosophers, for example Derek Parfit and others cited by him in *Reasons and*

Persons,[1] have shown, with the aid of intriguing thought experiments, that it is by no means obvious what it means when we say that we are the same person through time. Psychologists such as Bruce Hood have approached the same problem from other directions. This is no place for a philosophical disquisition, so I will content myself with the observation that continuity of memory makes me *feel* as though my identity has remained continuous during my whole life, while I simultaneously feel incredulous that I am the same person as the young book-despoiler and the young empathy-failure.

I was also a games-failure, but the school had a squash court and I became obsessed with squash. I didn't really enjoy trying to win against an opponent. I just liked knocking the ball against the wall by myself, seeing how long I could keep going. I had squash-withdrawal symptoms during the school holidays – missed the echoing sound as ball hit wall, and the smell of black rubber – and I kept dreaming of ways in which I might improvise a squash court somewhere on the farm, perhaps in a deserted pig sty.

Back at Chafyn Grove I would watch games of squash from the gallery, waiting for the game to end so I could slip down and practise by myself. One day – I must have been about eleven – there was a master in the gallery with me. He pulled me onto his knee and put his hand inside my shorts. He did no more than have a little feel, but it was extremely disagreeable (the cremasteric reflex is not painful, but in a skin-crawling, creepy way it is almost worse than painful) as well as embarrassing. As soon as I could wriggle off his lap, I ran to tell my friends, many of whom had had the same experience with him. I don't think he did any of us any lasting damage, but some years later he killed himself. The atmosphere at morning prayers told us that something was up even before Gallows made his grim

---

[1] Derek Parfit, *Reasons and Persons* (Oxford, Oxford University Press, 1984).

announcement, and one of the woman teachers was crying. Many years later in Oxford, a large bishop sat next to me at high table in New College. I recognized his name. He had been the (ah me, much smaller then) curate at St Mark's church, to which Chafyn Grove marched in crocodile for Matins every Sunday, and he was evidently in touch with the gossip. He told me that the same woman teacher had been hopelessly in love with the paedophile master who had killed himself. None of us had ever guessed.

While Sunday morning service was in St Mark's, every weekday morning and every evening we had prayers in the school chapel. Gallows was extremely religious. I mean really religious, not token-religious: he truly believed all that stuff, unlike many educators (and even clergymen) who pretend to do so out of duty, and politicians who pretend to do so because they are under the (I suspect exaggerated) impression that it wins votes. Gallows usually referred to God as 'the King' (he pronounced it 'Keeng', surprisingly because his speech was otherwise standard English 'received pronunciation'). I think that, when I was very young, this led to a certain confusion in my mind. I must have been aware that King George VI was not really God, but there was a certain almost synaesthetic confusion in my mind between royalty and godhead. This persisted after George VI's death into the coronation of his daughter, when Gallows instilled in us a deep reverence for ceremonial nonsense such as anointing with holy oil. I can still conjure an echo of the same reverence when I see a 1953 coronation mug, or hear Handel's magnificent anthem 'Zadok the Priest', or Walton's 'Orb and Sceptre' march, or Elgar's 'Pomp and Circumstance'.

Every Sunday evening we had a sermon. Gallows and Slush took turns to preach it, Gallows in his Cambridge MA gown with white hood, Slush in his Oxford MA gown with red hood. One extraordinary sermon sticks in my memory. I can't remember who preached it. Whoever it was told the story of a squad of soldiers drilling by a railway line. At one point the drill sergeant's attention was distracted and he failed to shout 'About turn!' So the soldiers continued to march on – right into the path of an oncoming train. The story can't have been true, and I now think it also can't be true that – as I seem to remember from the sermon – we were supposed to admire the soldiers for their unquestioning obedience to military authority. Perhaps it is a failure of my memory. I certainly hope it is. Psychologists such as Elizabeth Loftus have shown that false memories can be indistinguishable from true ones, even when deliberately planted by unscrupulous therapists seeking, for example, to persuade distressed people that they *must* have been sexually abused as children.

One Sunday a junior master, a nice young man called Tom Stedman, was cajoled, obviously with the utmost reluctance, to do the preaching. He clearly hated it. I remember his frequent repetition of 'What's a heaven for?' It would have made more sense if I had realized – at the time, instead of years later – that it was a quotation from Browning. Another popular young master, Mr Jackson, had a fine tenor voice. He was persuaded one day to sing Handel's 'The Trumpet Shall Sound', which he also did with extreme reluctance, evidently realizing – correctly – that his art was wasted on us.

Also wasted on us were the occasional visiting lecturers and per-formers, although I suppose the fact that I have remembered them must say something. The ones that stick in my memory are R. Keith Jopp on 'It's there still' (archeology), Lady Hull playing the upright piano in the dining hall (Schumann's *Faschingsschwank*), somebody talking about Shackleton's Antarctic expeditions, somebody else showing flickering black and white films of athletes of the 1920s and 1930s, including Sydney Wooderson, and a trio of Irish troubadours who set up a little stage for themselves and sang 'I bought my fiddle for ninepence, and that is Irish too'. One lecturer talked about explosives. He fished out what he claimed was a stick of dynamite. Casually saying that if he dropped it the whole school would go sky high, he threw it up and caught it. Of course we believed him, gullible little naïfs that we were. How could we not believe him? He was an adult, and we were brought up to believe what we were told.

We didn't believe only adults. We were gullible too in the dormitory, where the resident yarn-spinner fooled us nightly. He told us King George VI was his uncle. The unfortunate king was held a prisoner in Buckingham Palace, from where he smuggled out desperate messages in code with a searchlight to our dormitory raconteur, his nephew. This young fantasist terrified us with stories of a horrible insect that would leap sideways from the wall onto your head, dig a neat round hole the size of a marble in your temple and bury a bag of poison in the hole to kill you. He told us, during a violent thunderstorm, that if you were struck by lightning you would be completely unaware of the fact for fifteen minutes. Your first inkling would be when blood started trickling out of both ears. Shortly after that you would be dead. We believed him and waited on tenterhooks after each lightning flash. Why? What reason did we have to think that he knew any more than we did? Was it even remotely plausible that you would feel nothing when you were struck by lightning, until fifteen minutes later? Once again, that sad lack of critical thinking. Shouldn't children be taught critical, sceptical thinking from an early age? Shouldn't we all be taught to doubt, to weigh up plausibility, to demand evidence?

Well, perhaps we should, but we weren't. On the contrary: if anything, gullibility was positively encouraged. Gallows was extremely keen that we should all be confirmed into the Church of England before we left the school, and almost all of us were. The only exceptions I can remember were the one boy who came from a Roman Catholic family (and went to a different church every Sunday in the envied company of the pretty Catholic under-matron), and the one precocious boy who struck awe into us by claiming to be an atheist – he called the Bible the Holy Drivel and we expected thunderbolts daily (his iconoclasm, if not his logic, carried over into his style of geometric proof: 'Triangle ABC *looks* isosceles, therefore . . .').

I signed up for confirmation with the rest of my cohort; and the vicar of St Mark's, Mr Higham, came to give us weekly confirmation classes in the school chapel. He was a handsome, silver-haired, avuncular figure and we went along with what he said. We didn't understand it, it didn't seem to make sense, and we thought this was because we were too young to understand. It is only with hindsight that I realize that it didn't make sense because, quite simply, there was no sense to make. It was all invented nonsense. I still have, and frequently have occasion to refer to, the bible I was given on my confirmation. This time it was the real thing, the King James translation, and I have some of the best bits in my head to this day, especially Ecclesiastes and the Song of Songs (not of Solomon, needless to say).

My mother has recently told me that Mr Galloway telephoned the parents individually to say how keen he was to have us confirmed. He said that thirteen was an impressionable age, and it was a good idea for children to be confirmed early, in order to give them a steady base in religion before they had to confront counter-influences at their public schools. Well, you can't say he wasn't honest in his designs on naive young minds.

I became intensely religious around the time I was confirmed. I priggishly upbraided my mother for not going to church. She took it very well and didn't tell me, as she should have, to take a running jump. I prayed every night, not kneeling at the bed but curled up in a foetal position inside it, in what I confided to myself was 'my own little corner with God'. I wanted (but never dared) to steal down into the chapel in the middle of the night and kneel at the altar where, I believed, an angel might appear to me in a vision. If I prayed hard enough, of course.

In my final term, when I was thirteen, Gallows made me a prefect. I don't know why this pleased me so much, but I walked on air for the whole term. Later in my life, when the head of my department at Oxford was knighted by the Queen, I attended his celebratory party. I asked a colleague whether our professor was pleased by the honour and received the memorable reply: 'Like a dog with three pricks, old boy.' That's pretty much how I felt on being made a prefect. Also on being accepted into the Railway Club.

The Railway Club was the main reason I had been pleased by my parents' decision to send me to Chafyn Grove. It was run by Mr K. O. Chetwood Aiken, not really a teacher except on the rare occasions when a boy would opt to learn German. A melancholy man with a long, sad face, his real love and apparently sole pastime was his Railway Room (although I recently learned, on Googling him, that he had been a known Cornish artist). One room in the school was set aside for him and he built there a magical simulacrum of the Great Western Railway, 0-gauge electric, with two terminuses called Paddington and Penzance and one station halfway between called Exeter. Each engine had a name, Susan or George for example, and the two dear little shunting engines were both called Boanerges (Bo One and Bo Two). Each station had a bank of switches, each switch activating its own portion of track, red switches for the Up line and blue for the Down. When a train arrived at Paddington, you had to uncouple it from the big engine that had pulled it, then drive one of the little shunting engines from its siding to move the train from the Up line to the Down, then send the engine to the turntable to turn it around, then couple it to the new front of the train and send it back along the Down line to Penzance, where the whole process would be reciprocated. I loved the smell of ozone that came from the electric sparks, and I adored working out which was the right combination of switches to flick on and off for each operation. I think the pleasure I got from it was similar to that which I later derived from programming computers, and also from soldering the connections in my one-valve radio set. Everybody wanted to get into the Railway Club, and all who did so doted on Mr Chetwood Aiken despite his lugubrious mien. With hindsight I think he may already have been very ill, for he died of cancer not long after I left. I don't know whether the Railway Room survived his death, but I think the school would have been mad to let it go.

Much as I enjoyed the Railway Club and being allowed to sashay uninvited through the door of the prefects' study, the time came when I had to move on to another school and start at the bottom again. When I was only three months old my father had put my name down for Marlborough, his old school, but was told that he was too late: I should have been put down at birth (how long before that sentence is quoted out of context?). Marlborough's snooty letter was quite hurtful to him as an old boy, but he put my name on the waiting list anyway, and when the time came I could have gone to Marlborough. Meanwhile, however, my father's thoughts had turned in another direction. He was impressed by the technical skills of the next-door gentleman farmer, Major Campbell, who had a well-equipped workshop and was an expert welder. My father naturally thought that I might become a farmer, and workshop skills give great advantages in that career (as I have recently learned from one of the most successful and certainly the most unconventionally enterprising of farmers I have ever met, the redoubtable and heroic George Scales).[1]

Major Campbell had acquired his expertise at his old school,

---

[1] http://old.richarddawkins.net/articles/2127-george-scales-war-hero-and-generous-friend-of-rdfrs.

Oundle in Northamptonshire. Oundle had the finest workshops of any school in the country, and its great headmaster from 1901 to 1922, F. W. Sanderson, had initiated a system whereby every boy spent one whole week of every term in the workshops, all normal school work suspended. Neither Marlborough nor any other school could boast anything like that. My parents therefore put my name down for Oundle, and I took the scholarship exam in my last term at Chafyn Grove. I didn't get a scholarship, but I did well enough to get a place, and Oundle was where I went, in 1954, aged thirteen.

I don't know, by the way, how much else Major Campbell had picked up during his time at Oundle. I presume his robust approach to recalcitrant underlings came rather from his days in the army. He caught one of his workers in petty theft, I think of a tool from his workshop, and fired him in somewhat literalistic terms: 'I'll give you fifty yards' start before you get both barrels.' Of course he wouldn't have carried out the threat, but it makes a good story and another fine illustration of the shifting moral *Zeitgeist*.

'AND YOUR ENGLISH SUMMER'S DONE'

O F COURSE, there was life beyond school. At Chafyn Grove, we longed for the end of every term, and our favourite hymn was the one we sang on the last day: 'God be with you till we meet again'. It ranked even higher than the stirringly martial missionary hymn that we also loved:

> Ho, my comrades! See the signal waving in the sky
> Reinforcements now appearing, victory is nigh.
> 'Hold the fort, for I am coming,' Jesus signals still.
> Wave the answer back to Heaven, 'By thy grace we will.'

We all went joyfully home for the holidays, some on the school train to London, some fetched by parents in their cars – in my case a battered old Land Rover which never caused me the embarrassment that snobbish boarding-school children are alleged to feel when their parents show up in anything less expensive than a Jaguar. I was proud of the ragged leaky-roofed old war-horse, in which my father had driven us crashing through the undergrowth on a dead straight compass course, on the child-delighting theory that there must have been a Roman road connecting two colinear stretches of dead straight highway on the well-thumbed Ordnance Survey map.

Very typical of my father, that kind of thing. Like his own father, he loved maps; and both loved keeping records. Weather records, for instance. Year after year my father filled notebooks with meticulously dated measurements of the daily maximum and minimum temperatures, and of rainfall – his enthusiasm only slightly dampened when we caught the dog peeing into the rain gauge. We had no way of knowing how many times dear Bunch had done this before and how many past rainfall records were similarly augmented.

My father always had an obsessive hobby on the go. It was usually one that would exercise his practical ingenuity, which was considerable, although he was of the scrap metal and red binder-twine rather than the Major Campbell lathe and welding-kit school of thought. The Royal Photographic Society elected him a Fellow for his beautiful 'dissolving' productions. These were carefully crafted sequences of colour slides, displayed by twin projectors working side by side in alternation, each slide artistically fading into the next, with musical and spoken accompaniment. Today it would all be done by computer, but in those days the fading in and fading out had to be achieved by iris diaphragms, inversely linked so that each opened as the other closed. My father fashioned cardboard iris diaphragms for the two projectors, coupled to each other by a fiendishly ingenious system of rubber bands and red string, activated by a wooden lever.

Family tradition changed 'dissolving' to 'drivelling', because that is how it had once been misread in a hastily scribbled note. We all became so used to calling his art-form 'drivelling' that we never thought of calling it anything else and the word lost its original meaning. On one occasion my father was giving a public presentation (one of many around that time) to a photographic club. It happened that this particular presentation was largely put together from earlier photographs, taken before he had begun his 'dissolving' hobby, and he began by explaining this to the audience. He had an endearingly halting and rambling style of delivery, and the audience warmed, in a somewhat bemused way, to his opening sentence: 'Er, I actually, I actually, er, these photographs mostly date, er, mostly date from before I started *drivelling* . . .'

His less than fluently accomplished style of speaking had earlier shown itself during his courtship of my mother, when he lovingly looked deep into her eyes and murmured, 'Your eyes are like . . . spongebags.' Bizarre as this sounds, I think I can make some sort of sense of it and it again has something to do with iris diaphragms. When seen end-on, a spongebag's drawstrings look a bit like the radiating lines which are an attractive feature of an eye's iris.

Another year, his hobby was making pendants for all his female relations, each one a sea-smoothed Cornish serpentine pebble bound with a leather thong. At yet another time in his life, his obsession was to design and build his own automated pasteurizer for the dairy, with flashing coloured signal lights and an overhead conveyance system for the churns, which provoked a lovely verse from one of his employees, Richard Adams (not the famous rabbit man), who managed the pigs:

With clouds of steam and lights that flash, the scheme is most
   giganto,
While churns take wings on nylon slings like fairies at the panto.

My father had a ceaselessly creative mind. While cultivating a field on his little grey Ferguson tractor, wearing his battered old KAR hat and singing psalms at the top of his voice ('Moab was my washpot': by the way, the fact that he sang psalms emphatically didn't mean he was religious), he had plenty of time to think. He calculated that all the time spent doubling back at the end of each row was wasted. So he devised an ingenious scheme for zigzagging diagonally across and along the field with *shallow* turning angles, such that the whole field could be covered twice, in little more than the time it would normally take to cover it once.

Ingenious on the tractor he may have been, but always sensible he was not. On one occasion the clutch on the tractor stuck down. Unable to get it out of gear, he lay down on the ground beneath the clutch to see why it was stuck, and eventually succeeded in freeing it. Now, if you lie down under the clutch of a tractor you'll find that you are also lying directly in front of the large left rear wheel. The tractor enthusiastically leapt into action and ran him over, and all I can say is it was a good thing it was a Ferguson and not one of today's giant tractors. The little Fergie went bowling triumphantly across the field, and Norman, my father's employee who was standing there, was too dumbstruck with horror to do anything about it. My father had to sit up and tell him to chase after the tractor to stop it. Poor Norman was also too shaken to drive him to hospital, so my father had to do that himself. He spent some time in hospital with his leg in traction, but apparently suffered no lasting damage. His stay in hospital had the beneficial side-effect of prompting him to give up smoking his pipe. He never went back to it, and its only legacy was hundreds of empty baccy tins bearing the slogan 'And assuredly this is a grand old rich tobacco' which he was still using decades later for keeping assorted screws, nuts and washers and the miscellaneous dirty old metal scraps in which he took such delight.

Under the influence of an evangelical agricultural author called F. Newman Turner, and also perhaps of his eccentric friend from Marlborough and Oxford days, Hugh Corley, my father was an early convert to organic farming, long before it became fashionable or patronized by princes. He never used inorganic fertilizers or weed-killers. His organic farming mentors also disapproved of combine harvesters, and our farm was too small to justify one anyway, so in the early days we harvested with an old binder. It clattered noisily across the field behind the little grey tractor, scissoring the wheat or barley in front and spitting out neatly tied sheaves behind (I marvelled at the clever mechanism for tying the knots). And then the real work began, because the sheaves had to be stooked. An army of us walked behind the binder picking up sheaves two at a time and stacking them against each other to make little wigwams (stooks), six sheaves to a stook. It was hard work, leaving our forearms scratched and grazed and sometimes bleeding, but it was satisfying and we slept well that night. My mother would bring jugs of draught cider (scrumpy) out to the fields for the stookers, and a warm feeling of good fellowship suffused the Hardyesque scene.

The purpose of stooking was to dry the crop, after which the sheaves were carted and tossed up onto a rick. As a boy, I was not strong enough with a pitchfork to toss a sheaf right to the top of a high rick, but I tried hard and I envied my father's strong arms and horny hands, the equal of any of his employees. Weeks later, a threshing machine would be hired and parked next to the rick. The sheaves were fed in by hand, the grain threshed out and the straw baled. The farm workers all joined in with goodwill, regardless of what their real jobs might have been – cowman or pigman or general handyman or whatever. Later we moved with the times and hired a neighbour's combine.

In an earlier chapter I said I was a secret reader who used to escape to my bedroom with a book instead of rushing around outside in all weathers in true Dawkins tradition. Secret reader I may have been, but I can't honestly pretend that my reading in the school holidays had much to do with philosophy or the meaning of life or other such deep questions. It was pretty standard juvenile fiction: *Billy Bunter*, *Just William*, *Biggles*, *Bulldog Drummond*, Percy F. Westerman, *The Scarlet Pimpernel*, *Treasure Island*. For some reason my family disapproved of Enid Blyton and discouraged me from reading her. My Uncle Colyear gave me the Arthur Ransome books in succession, but I never really got on with them. I think I found them too girly, which was silly of me. Richmal Crompton's *William* has, I still think, genuine literary merit, with irony that can appeal to an adult as well as a child. And even the *Billy Bunter* books, though so formulaically written they might almost have been composed by computer, have pretensions to literary allusion in such phrases as: 'Like Moses of old, he looked this way and that, and saw no man' or 'Like a podgy Peri at the gates of Paradise'. *Bulldog Drummond* plumbs depths of jingoistic and racist bigotry which unmistakably label its era but passed over my naive young head. My maternal grandparents had a copy of *Gone with the Wind*, which I reread avidly on more than one summer holiday, never really noticing the paternalistic racism until I was older.

Family life at Over Norton was about as happy as family life gets. My parents were a united couple who celebrated their seventieth wedding anniversary together shortly before my father died in December 2010 aged 95. We were not a particularly rich family, but we weren't poor either. We had no central heating and no television, although the latter was from choice more than poverty. The family car was the dirty old Land Rover I mentioned or a cream van, neither of them luxurious but they did the job. Sarah's and my schools were expensive, and my parents surely had to skimp in other areas of life to send us there. Our childhood holidays were not in posh hotels on the Côte d'Azur but in army surplus tents in Wales, pelted with rain. On those camping trips we washed in an ex-Burma-Forest-Department canvas bath, warmed by the camp fire on which we also cooked our meals. Sarah and I, in our tent, heard our father sitting in his bath with his feet outside it, meditatively ruminating to himself, 'Well, I've never had me bath in me boots before.'

For three of my most formative years in my early teens I had the equivalent of an elder brother. Our great friends from Africa, Dick and Margaret Kettlewell, had stayed on in Nyasaland. Dick had at an unusually young age become Director of Agriculture, and distinguished himself in the job so resoundingly that he later became Minister of Lands and Mines in the provisional government on the way to full independence. When their son Michael, a playmate of mine in our very early days, turned thirteen he started as a boarder at Sherborne School in England; and, as with my father a generation before, the question arose of where he should go in the school holidays. I was delighted when he came to us. The age gap was only just over a year, and we did everything together: swimming in the freezing cold stream in the valley; indoor pursuits like chemistry sets, Meccano,[1] ping-pong, canasta, badminton, miniature snooker, various childish concoctions and recipes for making beetroot wine, or detergents, or vitamin pills. With Sarah, we had a junior farming enterprise called The Gaffers. My father gave us a litter of piglets, which we called The Barrels. We fed them daily and were wholly responsible for looking after them. Mike and I have remained life-long friends. Indeed, he is now my brother-in-law and the grandfather of most of my young relations.

There is a downside to having an elder brother in your formative years, however. It can mean that whenever you do anything, he actually does the operation and you pass him the instruments (since Mike later became a distinguished surgeon, the metaphor is not unsuitable). My Uncle Bill had a lifelong reputation for being 'no good with his hands', whereas my father had the opposite repu-tation, and it was probably for the same reason. The younger brother is apt to be the apprentice, never the master craftsman. The elder brother tends to be the decision-maker, the younger brother the follower, and early habits stick. Unlike my Uncle Bill, I didn't culti-vate a reputation for being no good with my hands. Nevertheless I was – and now am – no good. Mike did everything, with me as

---

[1] American: Erector Set.

superfluous assistant, and my father probably looked forward to my imminent exposure to the famous workshops of Oundle, which should set me belatedly in the footsteps of Major Campbell. But those workshops, as we shall see, proved to be a disappointment.

I was probably a disappointment as a naturalist, too, despite the rare privilege of spending a day with the young David Attenborough, when we were both guests of my Uncle Bill and Aunt Diana. Already famous but not yet a household name, he had been their guest on a filming expedition up-country in Sierra Leone, and they remained friends. When Bill and Diana moved to England and I happened to be staying with them, David brought his son Robert to visit, and had us children wading all day through ditches and ponds with fishing nets and jam jars on strings. I've forgotten what we were seeking – newts or tadpoles or dragonfly larvae, I expect – but the day itself was never to be forgotten. Even that experience with the world's most charismatic zoologist, however, wasn't enough to turn me into the child naturalist that both my parents had been. Oundle beckoned.

THE SPIRE BY
THE NENE

> *By* the boys, *for* the boys. The boys know best.
> Leave it to them to pick the rotters out
> With that rough justice decent schoolboys know.

John Betjeman, Summoned by Bells

I GOT the English public school experience too late – thank goodness – for the real cruelties of the John Betjeman era. But it was quite tough enough. There were ludicrous rules, invented 'by the boys for the boys'. The number of buttons you were allowed to undo on your jacket was strictly laid down according to seniority, and strictly enforced. Below a certain seniority level, you had to carry your books with a straight arm. Why? The masters must have known this sort of thing was going on, yet they did nothing to stop it.

The fagging system was still going strong, although happily it no longer is. (Note to American readers: this doesn't mean what you think. In British English, a 'faggot' is not a homosexual but a bundle of sticks or a rather nasty meatball. And 'fag' means cigarette or boring task or – as in this case – schoolboy slave.) Each house prefect at Oundle chose one of the new boys as his personal slave or fag. I was chosen by the deputy Head of House, known as Jitters because he had a tremor. He was kind to me, but I still had to do his every bidding. I had to clean his shoes, polish the brasses of his Cadet Corps uniform and make toast for him at teatime every day on a paraffin pressure stove in his study. I had to be ready to run errands for him at any time.

Not that fags were totally immune to sexual importuning. On four separate occasions I had to fend off nocturnal visits to my bed from senior boys much larger and stronger than I was. I suspect that they were driven by neither homosexuality nor paedophilia in the normal sense of the outside world, but by the simple fact that there were no girls. Pre-pubescent boys can be pretty in a girlish sort of way, and I was. There was also folklore, rife throughout the school, of boys having 'crushes' on other boys with girlish appeal. Once again, I was the victim of many such rumours, whose only real damage was the – considerable – time they wasted in idle gossip.

Many things about Oundle were intimidating after Chafyn Grove. In the Great Hall for morning prayers on my first day, new boys were yet to be assigned places and we had to find empty chairs where we could. I found one and timidly asked the big boy next to it whether it was taken: 'Not as far as I can observe' was his icily polite reply, and I felt crushed very small. After the treble chorus and foot-pumped harmonium of Chafyn Grove, Oundle's deep bellowing of 'New every morning is the love', accompanied by the massive, thundering organ, was alarming. The stooping headmaster in his black MA gown, Gus Stainforth, was formidable in a different way from Gallows. In nasal tones he exhorted us to 'break the back of the term's work' by the third week: I wasn't sure how you set about breaking the back of anything, let alone a term's work.

My form master in 4B1, Snappy Priestman, was a gentle man, cultivated, kind and civilized except when he (very occasionally) lost his temper. Even then, there was something oddly gentlemanly about the way he did it. In one of his lessons he caught a boy misbehaving. After a lull when nothing happened, he began to give us verbal warning of his escalating internal fury, speaking quite calmly as an objective observer of his own internal state.

> Oh dear. I can't hold it. I'm going to lose my temper. Get down below your desks. I'm warning you. It's coming. Get down below your desks.

As his voice rose in a steady crescendo he was becoming increasingly red in the face, and he finally picked up everything within reach – chalk, inkpots, books, wood-backed blackboard erasers – and hurled them, with the utmost ferocity, towards the miscreant. Next day he was charm itself, apologizing briefly but graciously to the same boy. He was a kind gentleman provoked beyond endurance – as who would not be in his profession? Who would not be in mine, for that matter?

Snappy had us reading Shakespeare and assisted my first appreciation of that sublime genius. We did *Henry IV* (both parts) and *Henry V*, and he himself played the dying Henry IV, chiding Hal for having taken the crown prematurely: 'Oh my son. God put it in thy mind to take it hence, that thou mightst win the more thy father's love, pleading so wisely in excuse of it.' He asked for a volunteer who could do Welsh (Williams) and Irish (Rumary: 'Oh, Rumary, you are a treasure.') Snappy read us Kipling, putting on a creditable Scottish accent for the hymn of the Chief Engineer, M'Andrew (that really is Kipling's spelling). The hauntingly rhythmical opening verse of 'The Long Trail' put me sadly in mind of the ricks of Over Norton and the 'all is safely gathered in' satisfaction of early autumn (please read it aloud to get the Kipling rhythm).

> There's a whisper down the field where the year has shot her yield,
> And the ricks stand grey to the sun,
> Singing: 'Over then, come over, for the bee has quit the clover,
> And your English summer's done.'

121

And, right on cue for the mellow fruitfulness, Mr Priestman read us Keats.

Our mathematics master that same year, Frout, was prone to dizzy fits. Once, before he arrived in the classroom, I seem to remember that we set all the lights swinging from the ceiling. Then when he came in we swayed in unison with them. I don't recall what happened next. Maybe remorse has blocked out the memory. Or maybe it is a false memory based on a schoolboy folk legend about what others had earlier done to him. Either way, I now see it as yet another example of the lamentable cruelty of children – a recurring theme of my schoolday recollections.

We didn't always get it our own way. One time the 4B1 physics master, Bufty, was ill and the class was taken instead by the senior science master, Bunjy. Having ascertained that we had reached Boyle's Law in our curriculum, he proceeded to teach us, labelling us with numbers in place of our names, which he had no time to learn. Small, stooping, old, and more short-sighted than anybody I have ever encountered before or since, he was, we thought, easy meat for ragging. He seemed scarcely to notice our insolence. We were wrong. Hypermyopic he may have been, but he noticed. At the end of the lesson, Bunjy quietly announced that he was keeping us all in detention that very afternoon. Crestfallen, we returned in the afternoon and were instructed to write on a clean page in our notebooks: 'Extra Lesson for Form 4B1. Object of the Lesson: To teach 4B1 good manners and Boyle's Law'. I am confident that this is not a false memory and I, for one, have never forgotten Boyle's Law.

One of our masters – the only one we were allowed to call by his nickname – was prone to fall in love with the prettier boys. He never, as far as we knew, went any further than to put an arm around them in class and make suggestive remarks, but nowadays that would probably be enough to land him in terrible trouble with the police – and tabloid-inflamed vigilantes.

Like most schools of its type, Oundle was divided into houses. Each boy lived in, and dined in, one of eleven houses, and his house commanded his loyalty in all competitive fields of endeavour. Mine was Laundimer. I don't know what the others were like inside because we were discouraged from visiting other houses, but I suspect that they were all much the same. Interestingly, however, our minds tended to see each house as having a 'personality', and we unconsciously grafted that personality onto individual boys in the house concerned. These house 'personalities' were so nebulous that I cannot find it in me to attempt a description of any one of them. It was just something one 'felt', subjectively. I suspect that this observation represents, in a somewhat more innocent form than many prevalent in the wider world, that 'tribal' human impulse that lies behind much that is more sinister, such as racial prejudice and sectarian bigotry. I'm talking about the human tendency to identify individuals with a group to which they belong, rather than seeing them as individuals in their own right. Experimental psychologists have shown that this happens even when individuals are allocated to groups at random in the first place and labelled with badges as arbitrary as T-shirts of different colours.

As a particular illustration of the effect – actually rather an agreeable one in this case – there was a single boy of African ancestry at Oundle when I was there. It is my impression that he suffered no racial prejudice whatsoever at that time, possibly because, being the only black boy, he was not identified with a racial group within the school. But he was identified with the house to which he belonged. Along with his contemporaries at Laxton House, we saw him not as noticeably black at all but as 'one of the Laxton crowd', with a similar personality to others in Laxton. In hindsight, I doubt that there was any identifiable personality trait that could reasonably be associated with Laxton or any other house. My observation relates not at all to the reality of life at Oundle but to a general characteristic of human psychology, the tendency to see individuals as badged with a group label.

My reason for choosing Laundimer as my house was a rumour, which proved ill-founded, that it was one of the few houses that lacked the tradition of an initiation ceremony (what American college students call 'hazing'). As it turned out, we did have to stand on the table and sing a song. In my piping treble I sang one of my father's songs:

> Oh the sun was shining, shining brightly
> Shining as it never shone before – shone before.
> Oh the sun was shining so brightly,
> When we left the baby on the shore.
>
> Yes we left the baby on the shore.
> It's a thing that we've never done before – done before.
> When you see the mother, tell her gently
> That we left the baby on the shore.

Singing this was an ordeal, but not as bad, in the event, as I had feared.

I didn't see much individual bullying at Oundle, but there was a kind of formalized bullying which afflicted every new boy for one week in his first term or two, at least in Laundimer, and I think much the same happened in the other houses. This was the dread week when he was 'bell boy'. In your week as bell boy you were responsible for everything, and you were to blame if anything went wrong – which it usually did. You had to light the fire and make sure it didn't go out. On Saturday during your week of ordeal as all-purpose scapegoat, you had to go round all the studies taking orders for Sunday newspapers, and collecting money for them. Then, on Sunday morning, you had to get up very early, walk to the far end of the town to buy the newspapers, then carry them back and distribute them to all the studies. Your most publicly noticeable function was to ring the bell at exactly the right time to signal each of many deadlines throughout the day: getting up time, mealtimes, bedtime and so on. That meant you had to have a very accurate watch. By the end of my week as bell boy, I had got the hang of it, but the first day was a disaster. For some reason, I hadn't grasped that the five-minute warning bell had to be rung *exactly* five minutes before the breakfast gong. Many of the senior boys were in the habit of getting out of bed five minutes to the dot before the gong rang, and five minutes is not long to wash and dress so the timing was crucial. On my first day as bell boy I rang the five-minute bell, then strolled across to belabour the gong about half a minute later. Consternation was rife, and angry ridicule ensued.

The duties of bell boy and fag were such that it is a wonder we new boys got any work done at all, let alone succeeded in 'breaking the back of the term's work'. Fagging has now been abolished, I think in all English schools. But I remain at a loss as to why it was ever permitted in the first place, and why it lasted as long as it did. In the nineteenth century there was a weird belief that it had some kind of educational value. Perhaps its long persistence had something to do with the 'I went through it in my time so why shouldn't you?' mentality – a mentality that is still, incidentally, the bane of many a junior doctor's life in Britain.

Not entirely surprisingly, my stammer resurfaced in my early terms at Oundle. I had trouble with hard consonants like 'D' and 'T' and it was unfortunate that my surname begins with one of them, for it was often necessary to enunciate it. When we had tests in class, we had to tick our correct answers, count the ticks and then shout out the tally, out of ten, for the master to record in his book. When I got ten out of ten, I used to call out 'nine' because it was so much easier to say than 't-t-t-ten'. In the army Cadet Corps we were to be inspected by a visiting general. One by one we would have to march out from the ranks, stamp to attention in front of him, shout our name, salute, smart about turn and march back. 'Cadet Dawkins, sir!' I dreaded it. I had sleepless nights about it. It was fine to practise by myself, but when I had to shout it out in front of the whole parade? 'Cadet D-d-d-d-d . . .' In the event, it passed off all right, with just a long, hesitating pause before the D.

The Cadet Corps was not quite compulsory. You could get out of it if you joined the Boy Scouts. Or the other way out was to spend the time tilling the land with Boggy Cartwright. In a previous book I described Mr Cartwright as 'a remarkable, bushy-browed man, who called a spade a spade and was seldom seen without one'. Although paid to teach us German, what he actually taught us, in a slow, rural accent, was a kind of earthy, agricultural eco-wisdom. His blackboard permanently had the word 'Ecology' written on it and if anybody erased it when he wasn't looking he promptly rewrote it without saying a word. When writing German on the blackboard, if a sentence threatened to overwrite 'Ecology', he would cause the German sentence to flow around and over it. He once caught a boy reading P. G. Wodehouse and furiously tore the book clean in two. He had evidently bought into the calumny – assiduously fostered by Cassandra of the *Daily Mirror* – that Wodehouse had been a German collaborator during the war, on a par with Lord Haw-Haw or – the American equivalent – Tokyo Rose. But Mr Cartwright had the story even more garbled than Cassandra's slander. 'Wodehouse once had the opportunity to kick a German colonel downstairs, and he didn't take it.' That makes him sound like an angry man. He really wasn't, except under extreme provocation, which, bizarrely, P. G. Wodehouse (he said 'Woadhouse' instead of the correct 'Woodhouse') seems to have constituted. He was just a wonderfully original character, ahead of his time in his ecological eccentricity, slow-spoken and literally down to earth.

I was not enterprising enough to get out of the Cadet Corps by either of the two escape routes. I was probably too influenced by my peers – which actually was the story of my life at Oundle. Eventually I got out of the worst parts of army training by joining the band, playing first the clarinet and then the saxophone, conducted by a bandsman NCO: 'Right, we'll go from the *very commencement* of the 'ole march.' Of course, being in the band didn't get us out of the weekly duty of polishing our army boots, blanco-ing our belts and shining our brasses with Duraglit or Brasso. And we had to go to army camp once a year, living in the barracks of some regiment or other, going on long route marches and fighting mock battles with blank ammunition in our antiquated Lee-Enfield rifles. We also fired live rounds at targets, and one boy in my platoon accidentally shot the adjutant in the fleshy part of the leg. He fell to the ground and immediately lit a cigarette, while we witnesses, still on the ground with our Bren guns, felt very queasy.

On one expedition to the Leicester barracks we were exposed to a real sergeant major, the genuine article complete with huge, waxed ginger moustache. He would bellow, 'Seeerloooooope ARMS' or 'Ordeeeeeer ARMS', the first word in each case being a bass and pro-longed bellow, while the second word was a staccato – and absurdly high – soprano shriek. We suppressed our laughter into terrified snorts, in the manner of Pontius Pilate's soldiers in the Monty Python 'Biggus Dickus' scene.

We had to pass an examination called Certificate A, which involved rote learning of army knowledge: an exercise clearly designed to suppress anything remotely resembling intelligence or initiative – commodities not valued in the ranks of general infantry. 'How many kinds of trees do we have in the army?' The correct answer was three: Fir, Poplar and Bushy Top (the poet Henry Reed picked up on this point, but our drill sergeants would not have appreciated his satire).

Peer pressure among schoolchildren is notoriously strong. I and many of my companions were abject victims of it. Our dominant motivation for doing anything was peer pressure. We wanted to be accepted by our fellows, especially the influential natural leaders among us; and the ethos of my peers was – until my last year at Oundle – anti-intellectual. You had to pretend to be working less hard than you actually were. Native ability was respected; hard work was not. It was the same on the sports field. Sportsmen were admired more than scholars in any case. But if you could achieve sporting brilliance without training, so much the better. Why is native ability more admired than hard graft? Shouldn't it be the other way around? Evolutionary psychologists might have interesting things to say on the question.

But such missed opportunities! There were all sorts of exciting clubs and societies, any of which I could have joined with benefit. There was an observatory with a telescope – perhaps the gift of an old boy – and I never went near it. Why not? I would be enthralled to do so now, to be instructed by a knowledgeable astronomer with a real telescope that I didn't have to set up myself. I sometimes think schooldays are too good to be wasted on teenagers. Perhaps devoted teachers, instead of casting their pearls before piglets, should be given the opportunity to teach pupils old enough to appreciate their beauty.

For me at Oundle, the biggest missed opportunity of all lay in the workshops, which were my father's main reason for sending me to the school in the first place. It wasn't entirely my fault. Sanderson's unique innovation of a compulsory week in the workshops was still in full swing, and the workshops were superbly well equipped. We learned how to use lathes, milling machines and other advanced machine tools which we would be unlikely to meet in the big world outside. What we did not learn was precisely what my father was so good at: improvising, designing, making do and coping, knocking things up from what was available – in his case, mostly red binder twine and dirty old bits of iron.

The first thing we made in the Oundle workshops was a 'marking gauge'. We weren't even told what a marking gauge was. We copied exactly what the instructors told us to do. We made a wooden pattern for the metal object we were trying to make. We took it into the foundry and made a mould of our wooden pattern by pounding sticky sand around it. We donned protective goggles and assisted in pouring molten aluminium from a glowing crucible into the mould. We disinterred the cooled metal from the sand and took it to the metalwork shop to file it, drill it and finish it. And we took home our finished marking gauge, still with no idea what a marking gauge was and having used no initiative or creativity of any kind. We might as well have been workers in a mass production factory.

And part of the problem may indeed have been that the instructors were not teachers but were recruited – I'm guessing – from the ranks of factory floor foremen. They taught us not how to develop skills in general, but how to do particular things. I met the problem again when I took professional driving lessons in the town of Banbury. I was taught how to reverse round a particular corner in Banbury, which happened to be the favourite corner the examiner headed for when testing that particular skill: 'Wait till that lamppost is level with the back window, then swing hard around.'

The one exception in the Oundle workshops, the one partial upholder of the Sanderson tradition for me, was an old retired blacksmith who manned a little forge in a corner of the metal shop. I hived myself off from the 'factory floor' and apprenticed myself to this kindly, bespectacled little old man. He taught me the traditional arts of the smith, plus acetylene welding, and my mother still has the poker I made, sitting in its scrolled stand. Even with the old smith, however, I pretty much did exactly what I was told, rather than exercising much creative resourcefulness.

A bad workman blames his tools – and his instructors. What was definitely my own fault is that I never went near the workshops except during the prescribed week. I didn't seize the opportunity to go in the evenings and make things to my own design. Just as I didn't go to the observatory to look at the stars. Mostly I wasted my spare time in the same way my colleagues did, lazing around, making toast on a Primus stove and listening to Elvis Presley. Plus, in my case, tootling on musical instruments rather than playing real music. Such a waste of first-class, expensively bought opportunities is little short of tragic. Once again, is school too good for teenagers?

I did, however, join the beekeeping club, run by Ioan Thomas, Oundle's inspiring young zoology master, and the smell of beeswax and smoke still evokes happy memories. Happy in spite of the fact that I was quite frequently stung. On one such occasion (I am mildly proud to report) I didn't brush the bee off my hand but carefully watched as she slowly waltzed round and round on my hand, 'unscrewing' her sting from my skin. The stings of bees, unlike those of wasps, are barbed. When a bee stings a mammal, the barbs cause the sting to stick in the skin. When you brush the bee off, the sting stays behind and tears out some of the bee's vital organs. From an evolutionary point of view, the individual worker bee is behaving altruistically, sacrificing her life as a kamikaze fighter for the benefit of the hive (strictly speaking, for the benefit of the genes that programmed her to do it, in the form of copies in queens and males). While she goes off to die, her sting remains in the victim, the poison gland still pumping venom and therefore acting as a more effective deterrent to the putative hive-raider. This makes perfect evolutionary sense, and I'll return to the theme in the chapter on *The Selfish Gene*. Given that she is sterile, the worker bee has no chance of passing on copies of her genes via offspring, so instead she works to pass them on via the queen and other non-sterile members of the hive. When I let my worker unscrew herself from my hand I was behaving altruistically towards her – but my motivation was mostly curiosity: I wanted to watch at first hand the procedure I had heard about from Mr Thomas.

I've mentioned Ioan Thomas in previous publications. My very first lesson with him, at the age of fourteen, was inspirational. I don't remember the details, but it conveyed the kind of atmosphere I was later to strive for in *Unweaving the Rainbow*: what I would now call 'science as the poetry of reality'. He had come to Oundle as a very young teacher because of his admiration for Sanderson, although he was too young to have met that old headmaster. He did meet Sanderson's successor, Kenneth Fisher, and told a story showing that something of the spirit of Sanderson had lived on. I retold the tale in my inaugural Oundle Lecture, given in 2002.

Kenneth Fisher was chairing a staff meeting when there was a timid knock on the door and a small boy came in: 'Please, sir, there are Black Terns down by the river.' 'This can wait,' said Fisher decisively to the assembled committee. He rose from the Chair, seized his binoculars from the door and cycled off in the company of the small ornithologist, and – one can't help imagining – with the benign, ruddy-faced ghost of Sanderson beaming in their wake. Now that's education – and to hell with your league table statistics, your fact-stuffed syllabuses and your endless roster of exams . . .

Some 35 years after Sanderson's death, I recall a lesson about *Hydra*, a small denizen of still freshwater. Mr Thomas asked one of us: 'What animal eats *Hydra*?' The boy made a guess. Non-committally, Mr Thomas turned to the next boy, asking him the same question. He went right round the entire class, with increasing excitement asking each one of us by name, 'What animal eats *Hydra*? What animal eats *Hydra*?' And one by one we guessed. By the time he had reached the last boy, we were agog for the true answer. 'Sir, sir, what animal does eat *Hydra*?' Mr Thomas waited until there was a pin-dropping silence. Then he spoke, slowly and distinctly, pausing between each word.

'I don't know . . .' *(Crescendo)* 'I don't know . . .' *(Molto crescendo)* 'And I don't think Mr Coulson knows either.' *(Fortissimo)* 'Mr Coulson! Mr Coulson!'

He flung open the door to the next classroom and dramatically interrupted his senior colleague's lesson, bringing him into our room. 'Mr Coulson, do you know what animal eats *Hydra*?' Whether some wink passed between them I couldn't say, but Mr Coulson played his part well: he didn't know. Again the fatherly shade of Sanderson chuckled in the corner, and none of us will have forgotten that lesson. What matters is not the facts but how you discover and think about them: education in the true sense, very different from today's assessment-mad exam culture.

Those two occasions, when I fancifully invoked the ghost of a long-dead headmaster, have been held up as showing that I must be in some sense a supernaturalist. Of course they show nothing of the kind. Such imagery should perhaps be called poetic. It is legitimate so long as it clearly is understood to be non-literal. I hope the context of those two quotations is sufficiently clear to obviate misunderstanding. Problems arise when (especially) theologians use such metaphorical language without realizing that that is what they are doing, and without even realizing that there is a distinction between metaphor and reality – saying something like: 'It is not important whether Jesus really fed the five thousand. What matters is what the *idea* of the story *means* to us.' Actually it is important, because millions of devout people do believe the Bible is literally true. I hope and trust that no reader thinks I believe Sanderson really was standing in the corner beaming at Mr Thomas's lesson.

Our lesson on *Hydra* was the scene of a slightly embarrassing story, but I should tell it as it might be revealing. Mr Thomas asked us whether any of us had seen *Hydra* before. I think I was the only boy to put his hand up. My father had an old brass microscope, and we had spent a lovely day a few years earlier looking at hugely magnified pond life: mostly crustaceans such as *Cyclops, Daphnia* and *Cypris*, but also *Hydra*. I had regarded the slowly waving, almost plant-like *Hydra* as rather dull compared with the crustaceans, leggy and vigorously kicking. *Hydra* was the least exciting memory of that memorable day, and I think I snobbishly looked down upon all the attention that Mr Thomas was giving to it in that lesson. So, when he asked me for more details of my previous encounter with *Hydra*, I said: 'I've seen all those sorts of animals.' To Mr Thomas, of course, *Cyclops, Daphnia* and *Cypris* were not at all the same sort of animal as *Hydra*, but to me they were because I had seen them all on the same day with my father, and so lumped them together. Mr Thomas probably suspected that I hadn't seen *Hydra* at all, and he cross-examined me closely. I am sorry to say that this had exactly the wrong effect on me. Perhaps I took his cross-questioning as some sort of slur on my father, who had introduced me to 'all those sorts of animals' and told me their Latin names. I obstinately dug in my toes and, instead of saying, clearly and unequivocally (and truthfully), that I had indeed seen *Hydra*, I persisted in my refusal to separate it from 'all those sorts of animals'. Embarrassing to recall. Revealing? Maybe, but I don't know of what. Perhaps it was connected with the fierce loyalty that I felt towards all things associated with my parents, whether it was Ferguson tractors ('Dirty old Fordson!') or Jersey cows ('Friesians don't give milk, they give water').

Mr Thomas having introduced me to beekeeping, I was able to carry on with the hobby in the school holidays when my father's eccentric old schoolfriend Hugh Corley gave me a hive. They were a wonderfully docile strain which literally never stung, and I used to work them without veil or gloves. Unfortunately they were later poisoned by insecticide wafting over from a neighbour's field. Mr

Corley, passionate organic farmer and early eco-warrior, was outraged and gave me another hive. Unfortunately these went to the opposite extreme – undoubtedly a genetic difference – and stung everything that moved. I didn't react badly to stings in those days. But I wonder whether those many stings in my boyhood sensitized me to stings in later life. I have been stung only twice as an adult, once in my forties and once in my fifties, and on both occasions I reacted strangely and in a way that never happened when I was an active beekeeper. The region around one eye swelled up hugely, so that I could scarcely see. Why the eye, given that the stings were respectively on hand and foot? And, especially, why only one eye?

Apart from beekeeping with Mr Thomas, I suppose my other mildly constructive spare-time occupation at Oundle was playing music. I spent many hours in the Music School, but even there I have to confess to massive wasting of opportunities. From my earliest childhood, musical instruments of any kind would draw me like a magnet, and I had to be dragged away from shops that had violins or trumpets or oboes in the window. Even today, if a string quartet or a jazz band has been engaged to play at a garden party or a wedding, I will neglect my social duties and hover around the musicians, watching their fingers and talking to them during the intervals about their instruments. I don't have perfect pitch like my first wife Marian, and my harmonic sense is poor, unlike that of my present wife Lalla, who can effortlessly improvise harmonious descants to any melody. But I do have a natural melodic ability, meaning that I can play a tune about as easily as I can sing it or whistle it. I'm sorry to say that one of my pastimes in the Music School was illicitly to pick up instruments that didn't belong to me and teach myself to play tunes on them. On one occasion I was caught playing 'When the Saints Go Marching In' on a rather expensive trombone belonging to a senior boy, and got into trouble because the trombone was later found to be damaged. I genuinely believe I didn't do the damage, but I was blamed (not by the owner himself, who was rather nice about it).

My facile melodic gift turned out to be a curse rather than a blessing, at least in a child as lazy as I was. Playing by ear was so easy for me that I neglected other important skills such as reading music or creative improvisation. It was worse than laziness. For a while I even snobbishly looked down upon musicians who 'needed' to read music. I thought improvisation was a superior skill. But it turned out that I was no good at improvising either. Invited to join the school jazz band, I soon discovered that, although I could play any tune faultlessly, I had absolutely no capacity to improvise upon it. I was very slovenly about practising scales. I have a very slight, partial excuse, which is that nobody ever explained to me what scales are for. With hindsight, as an adult scientist, I can piece the reason together. You play scales in order to become totally at home with every key, so that, once you've read the key signature at the beginning of the line, your fingers automatically and effortlessly feel their way into that key.

The hours I spent in the Music School are best described as tootling rather than playing. I did learn to read a score adequately with the clarinet and saxophone. But on the piano – where you are expected to play more than one note at a time – I was unbearably slow, like a child learning to read and laboriously spelling his way through the words letter by letter, rather than fluently reading whole sentences at a time. My kind piano teacher, Mr Davison, recognized my innate melodic ability and taught me some rudimentary rules for accompanying myself with left-hand chords. But though I quickly learned these, I could do them only in the keys of C major and A minor (minimizing the black notes), and my style of left-hand chord-thumping was pretty monotonous – although inexpert listeners were impressed by my ability to play instant requests.

I had a true and pure, though not very loud, singing voice as a treble, and was early recruited into the rather small and select Chancel Choir in the Oundle school chapel. I hugely enjoyed this; the regular rehearsal, under the Director of Music Mr Miller, was the high spot of my week. I think it was rather a good choir, up there with a typical English cathedral choir. And I can't resist adding that we sang without the affectation of the half-rolled 'r' – sounding more like a 'd' – which, at least to my prejudiced ear, spoils much choral singing: 'Maady was that mother mild / Jesus Cdist, her little child.' 'The dising of the sun / And the dunning of the deer / The playing of the meddy organ . . . ' By the way, while I'm doing my grumpy act, the fake Italian 'r' of John McCormack-vintage tenors is even worse: 'Seated one day at the Oregon . . .'

We performed an anthem every Sunday: Stanford or Brahms or Mozart or Parry or John Ireland, or earlier composers such as Tallis or Byrd or Boyce. We had no conductor, but two of the basses, facing each other in the back rows on the two sides of the chancel, performed the role by their miming head movements. One of these basses, C. E. S. Patrick, had a spellbindingly beautiful voice – probably the better for not being trained. I never spoke to him (one didn't meet senior boys in other houses), but I hero-worshipped him as the star of the Male Voice Choir, which performed under the direction of another gifted music master, Donald Payne, at school concerts. Unfortunately I was never invited to join the Male Voice Choir. When my voice broke, it dropped in quality as well as in pitch.

Oundle had a tradition – again founded by Sanderson – of involving the entire school in an annual oratorio. The choice of music was staggered in such a way that every boy would experience Handel's *Messiah* and Bach's B Minor Mass during his five years at the school. The intervening years offered a variety of works. My first term we did Bach's *Sleepers Wake* cantata and Haydn's Imperial

Mass, and I *loved* them, especially the Bach, with its slow chorale for the voices cunningly set against the leaping counterpoint melody in the orchestra. This was a magical experience, of a kind I had never known before. Every morning, for five minutes after prayers, the tall, thin figure of Mr Miller would stride briskly forward and rehearse the entire school, just a few pages at a time, until the big day came for the performance. Professional soloists arrived from London: glamorous soprano and contralto in long dresses, tenor and bass in immaculate tailcoats. Mr Miller treated them with great deference. Goodness knows what they thought of the throaty roar of the 'non-choir'. But none of the soloists, in my youthfully amateur opinion, could hold a candle to C. E. S. Patrick of the Male Voice Choir.

It is hard to convey the atmosphere of the English public school during the era that I experienced it. Lindsay Anderson captured it well in his film *If.* I'm not referring to the massacre at the end of the film, of course, and he exaggerated the beating. Maybe prefects with swagger sticks and embroidered waistcoats took a run at it in earlier, crueller eras, but I'm sure it didn't happen in my time. Actually, I never knew of anyone being caned at all while I was at Oundle and only recently heard (from a victim) that it did happen.

If also beautifully captured the burgeoning sexuality that surrounds pretty boys in a school that has no girls. The flashlight inspection of groins by the matron in the enormous starched hat was only slightly exaggerated in the film. Our inspection was done by the school doctor, who didn't peer as pruriently as the *If* matron. Nor did our mild doctor stalk the touchline of the rugby field like she did, screaming 'Fight! Fight! Fight!' But what Lindsay Anderson caught to perfection was the squalid conviviality of the studies where we mostly lived, worked, burnt toast, listened to jazz and Elvis, and fooled around. He caught the hysterical laughter that bonded teenage friends like wrestling puppies – not physical wrestling but verbal wrestling with odd, private languages and weird nicknames that grew and evolved term by term.

As an illustration of the weirdness of nickname evolution (and maybe of memetic mutation generally), one friend of mine was called 'Colonel', although there was nothing remotely military about his personality. *'Seen the Colonel anywhere?'* Here's the evolutionary history. Years earlier, an older boy, who had by now left the school, was said to have had a crush on my friend. That older boy's nickname was Shkin (corruption of Skin, and who knows where that came from – maybe some connection with foreskin, but that name would have evolved before I arrived). So my friend inherited the name Shkin from his erstwhile admirer. Shkin rhymes with Thynne, and at this point something akin to Cockney rhyming slang stepped in. There was a character in the BBC radio *Goon Show* called Colonel Grytte Pyppe Thynne. Hence my friend became Colonel Grytte Pyppe Shkin, later contracted to 'Colonel'. We loved the *Goon Show*, and would vie with each other to mimic (as did Prince Charles, who went to a similar school around the same time) the voices of the characters: Bluebottle, Eccles, Major Denis Bloodnok, Henry Crun, Count Jim Moriarty. And we gave each other Goon nicknames like 'Colonel' or 'Count'.

Some of the squalor would positively not be allowed by a health inspector today. After playing rugby we would have a 'shower'. My hypothesis is that at some time in the past it really had been a shower, and other houses in the school probably had proper showers still. But in Laundimer House, all that was left of the shower was the porcelain rectangular base, which we would fill with hot water. It was just big enough for two boys to sit in, face to face, with their knees up under their chins. We queued up to enter the 'shower' and by the time all fifteen rugby players had been through it the 'water' was not so much water as dilute mud. The odd thing is that I don't think we minded being in the last pair. It had the advantage that you could linger on in the warmth instead of rushing to let the queue go through. I don't remember minding the fact that I was bathing in the muddy bathwater of fourteen other people, any more than I minded getting in a very small bath with another naked male – both things that I would dislike intensely today. Another indication, I suppose, that we are not the same people we once were.

Oundle didn't really live up to my parents' expectations. The vaunted workshops were a failure, at least where I was concerned. There was too much adulation of the rugby team and too little prestige attached to intelligence or scholarship, or indeed any of the qualities that Sanderson fostered. But in my last year at least, my set of peers finally started valuing the mind. A bright young history master started a club called Colloquium for intellectual discussion among sixth-formers. I can't remember what happened at the meetings: maybe we even used to 'read a paper', like earnest under-graduates. Equally earnestly, outside the meetings we would evaluate each other's intelligence, in an atmosphere of po-faced snobbery not unlike that conjured by John Betjeman's couplet:

> Objectively our common room is like a small Athenian state . . .
> Except for Lewis: he's all right, but do you think he's quite first
> rate?

I and two friends in my house became militantly anti-religious in our last year, when we were seventeen. We refused to kneel down in chapel and sat with folded arms and closed lips, defiantly upright like proud, volcanic islands in the sea of bowed and mumbling heads. As you'd expect of Anglicans, the school authorities were very decent and never complained, even when I took to skipping chapel altogether. But here I need to go back and trace my loss of religious faith.

I had arrived at Oundle a confirmed Anglican, and I even went to Holy Communion a few times in my first year. I enjoyed getting up early and walking through the sunlit churchyard listening to the blackbirds and thrushes, and I basked in righteous hunger for breakfast afterwards. The poet Alfred Noyes (1880–1958) wrote: 'If ever I had any doubts about the fundamental realities of religion, they could always be dispelled by one memory – the light upon my father's face as he came back from early communion.' It's a spectacularly silly piece of reasoning for an adult, but it sums me up at the age of fourteen.

I'm happy to say it wasn't long before I reverted to earlier doubts, first planted at the age of about nine when I learned from my mother that Christianity was one of many religions and they contradicted each other. They couldn't all be right, so why believe the one in which, by sheer accident of birth, I happened to be brought up? At Oundle, after my brief phase of going to Communion, I gave up believing in everything that was particular about Christianity, and even became quite contemptuous of all particular religions. I was especially incensed by the hypocrisy of the 'General Confession' in which we mumbled in chorus that we were 'miserable offenders'. The very fact that the exact words were written down to be repeated the following week, and the week after and for the rest of our lives (and had been so repeated ever since 1662), sent a clear signal that we had no intention of being anything other than miserable offenders in the future. Indeed, the obsession with 'sin' and the Pauline belief that everybody is born in sin, inherited from Adam (whose embarrassing non-existence was unknown to St Paul), is one of the very nastiest aspects of Christianity.

But I retained a strong belief in some sort of unspecified creator, almost entirely because I was impressed by the beauty and apparent design of the living world, and – like so many others – I bamboozled myself into believing that the appearance of design demanded a designer. I blush to admit that I had not at that stage worked out the elementary fallacy of this argument, which is that any god capable of designing the universe would have needed a fair bit of designing himself. If you are going to allow yourself to conjure a designer out of thin air, why not apply the same indulgence to that which he is supposed to have designed, and cut out, so to speak, the middle man? In any case, of course, Darwin provided the magnificently powerful alternative to biological design which we now know to be true. Darwin's explanation had the huge advantage of starting from primeval simplicity and working up, by slow, gradual degrees, to the stunning complexity that pervades every living body.

But at the time the 'it's all so beautiful, there must have been a designer' argument swayed me. My faith was reinforced by, of all people, Elvis Presley, of whom I was a dizzily enthusiastic fan, like most of my friends. I bought his records as soon as they were released: 'Heartbreak Hotel', 'Hound Dog', 'Blue Moon', 'All Shook Up', 'Don't be Cruel', 'Baby I Don't Care' and many others. Their sound is irrevocably – it seems now so appropriate – linked in my mind with the faintly sulphurous smell of the ointment with which many of us battled our adolescent spots. I once embarrassed myself by singing 'Blue Suede Shoes' loudly at home, thinking I was alone in the house and not knowing that my father was in earshot. 'You can knock me down / Step on my face / Slander my name / All over the place.' To imitate Elvis properly in this song you have to rasp the words with a kind of venom, like a modern rap performer. It took my chagrined self a while to convince my father that I was not having some kind of fit, or suffering from Tourette's Syndrome.

So, I worshipped Elvis and I was a strong believer in a non-denominational creator god. And it all came together when I passed a shop window in my home town of Chipping Norton and saw an album called *Peace in the Valley* featuring a song called 'I Believe'. I was transfixed. Elvis was religious! In a frenzy of excitement I dived into the shop and bought it. Hurrying home, I slipped the record out of the sleeve and on to the turntable. I listened with delight – for my hero sang that every time he saw the wonders of the natural world around him, he felt his religious faith reinforced. My own sentiments exactly! This was surely a sign from heaven. Why I was surprised that Elvis was religious is now beyond me. He came from an uneducated working-class family in the American South. How could he *not* have been religious? Nevertheless I was surprised at the time, and I sort of half-believed that in this unexpected record Elvis was speaking personally to me, calling me to devote my life to telling people about the creator god – which I should be especially well qualified to do if I became a biologist like my father. This seemed to be my vocation, and the call came from none other than the semi-divine Elvis.

I am not proud of this period of religious frenzy, and I'm happy to say that it didn't last long. I became increasingly aware that Darwinian evolution was a powerfully available alternative to my creator god as an explanation of the beauty and apparent design of life. It was my father who first explained it to me but, to begin with, although I understood the principle, I didn't think it was a big enough theory to do the job. I was biased against it by reading Bernard Shaw's preface to *Back to Methuselah* in the school library. Shaw, in his eloquently muddled way, favoured Lamarckian (more purpose-driven) and hated Darwinian (more mechanistic) evolution, and I was swayed towards the muddle by the eloquence. I went through a period of doubting the power of natural selection to do the job required of it. But eventually a friend – one of the two, neither of them biologists, in whose company I later refused to kneel in chapel – persuaded me of the full force of Darwin's brilliant idea and I shed my last vestige of theistic credulity, probably at the age of about sixteen. It wasn't long then before I became strongly and militantly atheistic.

I said that the school authorities were decently Anglican about my refusal to kneel in chapel, and turned a blind eye. But that may not be quite true, at least not of two of them. The first was my English teacher at the time, Flossie Payne, familiar as an erect figure on his sit-up-and-beg bicycle with raised umbrella. Flossie publicly challenged me in class to explain why I was leading a rebellion against kneeling in chapel. I'm afraid I didn't give a good account of myself. Far from seizing the opportunity to lead my classmates in the same direction, I miserably stammered something about an English lesson not being the appropriate place to have the discussion, and retreated into my shell.

Second, I have only recently learned that my housemaster, Peter Ling (actually a nice man, if rather too conformist and conventional), telephoned Ioan Thomas, my zoology master, to voice his concern about me. In a recent letter to me, Mr Thomas reported that he warned Mr Ling that 'requiring someone like you to attend chapel twice a day on Sunday was doing you positive harm. The phone went down without comment.'

Mr Ling also summoned my parents for a heart-to-heart talk, over tea, about my rebellious behaviour in chapel. I knew nothing of this at the time and my mother has only just told me of the incident. Mr Ling asked my parents to try to persuade me to change my ways. My father said (approximately, by my mother's recollection): 'It is not our business to control him in that sort of way, that kind of thing is your problem, and I'm afraid I must decline your request.' My parents' attitude to the whole affair was that it wasn't important.

Mr Ling, as I said, was in his way a decent man. A contemporary and friend of mine in the same house recently told me the following nice story. He was illicitly up in a dormitory during the day, kissing one of the housemaids. The pair panicked when they heard a heavy tread on the stairs, and my friend hastily bundled the young woman up onto a window sill and drew the curtains to hide her standing shape. Mr Ling came into the room, and must have noticed that only one of the three windows had the curtains drawn. Even worse, my friend noticed, to his horror, that the girl's feet were clearly visible protruding under the curtain. He firmly believes that Mr Ling must have realized what was going on but pretended not to, perhaps on 'boys will be boys' grounds: 'What are you doing up in the dormitory at this hour?' 'Just came up to change my socks, sir.' 'Oh, well, hurry on down.' Good call on Mr Ling's part! The boy went on to become probably the most successful Old Oundelian of his generation, the knighted chief executive officer of one of the largest international corporations in the world, and a generous benefactor of the school, endowing, among other things, the Peter Ling Fellowship.

The headmaster of a large school is a remote and formidable figure. The stooping Gus Stainforth only taught me for one term – Divinity – and we were terrified of him. We read *The Pilgrim's Progress*, and then had to produce our own artist's impression of that rather unpleasant book. Halfway through his expected time at Oundle, Gus left to head his own old school, Wellington, and was succeeded at Oundle by Dick Knight, a large, athletic man who won our respect by his ability to hit a ball out of the ground (he had played cricket for Wiltshire) and by the way he sang with the 'non-choir' in the annual oratorio. He drove a big Rolls-Royce, 1920s vintage I would guess from its imposingly upright style – very different from the sleek purrers of later decades. He happened to be visiting Oxford on business at the same time as I and another boy were taking the Oxford entrance exam and being interviewed in our respective colleges of choice. When they heard this, Mr and Mrs Knight kindly offered us a lift back to Oundle in their ancient Rolls, and on the journey he discreetly raised the subject of my rebellion against Christianity. It was a revelation to talk to a decent, humane, intelligent Christian, embodying Anglicanism at its tolerant best. He seemed genuinely interested in my motives and not at all inclined to condemn. Years later, I was not surprised to learn from his obituary that, an outstanding classical scholar in his youth as well as a noted athlete, in retirement he took a degree in mathematics from the Open University. Sanderson would have loved him.

My father and grandfather had never contemplated any destination for me after Oundle other than Balliol College, Oxford. At the time, Balliol still retained its reputation as the foremost Oxford college, top of the examination league table and *alma mater* of a glittering list of distinguished old members: writers, scholars, statesmen, prime ministers and presidents all around the world. My parents went to see Ioan Thomas about my prospects. Mr Thomas was realistically frank: 'Well, he might just scrape into Oxford, but Balliol is probably aiming too high.'

Mr Thomas might doubt that I was good enough for Balliol but – great teacher that he was – he was determined that I should give it my best shot. He had me round regularly at his home in the evenings for extra tuition (unpaid, of course; he was that sort of teacher), and by some miracle he got me into Balliol. More importantly, that meant I got into Oxford. And insofar as anything was the making of me, Oxford was.

DREAMING SPIRES

'M R DAWKINS? Sign here, sir. I remember your three brothers, very fine winger one of them was. I don't suppose you play rugby, sir?'

'No, I'm afraid not, and, er, actually I never had any brothers. You must be thinking of my father and my two uncles.'

'Yes, sir, very fine young gentlemen, sign here please. You are on Staircase 11, Room 3, sharing with Mr Jones. Who's next?'

Well, that's approximately how the conversation went. I didn't write it down at the time. The Balliol College porter took the timeless view characteristic of his bowler-hatted profession. Young gentlemen might come and go, but the college goes on for ever. Indeed, it was to celebrate its 700th anniversary during my time there. Talking of that loyal and ancient bowler-hatted profession, I can't resist an anecdote more recently told me by the Head Porter of my present college, New College (well, it was new in 1379). An inexperienced new porter hadn't yet got the hang of the porters' incident book and what it was for. His entries in the log for his first night duty, at hourly intervals, consisted of (approximately; the details will be wrong):

8 p.m. Raining.

9 p.m. Still raining.

10 p.m. Raining harder.

11 p.m. Raining harder still. I could hear it banging on me bowler as I did me rounds.

Oxford, I should explain, is a federal university: a federation of thirty or so colleges, of which Balliol is one of three claiming to be the oldest. Except for the newer colleges, each one is built around a series of quadrangles. These beautiful old buildings mostly don't have horizontally running corridors like hotels or halls of residence, with rooms along a passageway: instead, there are lots of staircases leading off doors from the quadrangle, each staircase giving access to a number of rooms on three or four floors. Thus each room is known by a staircase number and a room number within its staircase. In order to visit a near neighbour, you'd probably have to go out into the quadrangle and then in at another staircase entrance. In my time there was a bathroom on every staircase, so we no longer had to go out into the cold in our dressing gowns. Nowadays, the rooms are more likely to have their own en-suite bathrooms, which my father would have called 'terribly molly' (soft, namby-pamby). I suspect that a large part of the motivation for installing them is to cater for the lucrative conference trade, which all the Oxford and Cambridge colleges ply out of termtime.

The colleges at both Oxford and Cambridge are financially autonomous self-governing institutions, some of them, such as St John's, Oxford, and Trinity, Cambridge, very wealthy. Trinity, by the way, is outstandingly rich in achievement as well as money. This one Cambridge college can boast more Nobel Prizes than any single *country* in the world except the USA, Britain (obviously), Germany and France. The University of Oxford can make the same proud

My grandmother Enid with her dog Susan (*left*) in the garden of The Hoppet, where my parents first met. On the eve of war they were married (*above*) from Water Hall, seen below with my mother's younger sister Diana in the garden.

My mother made the journey to Africa in a series of low hops in the flying boat *Cassiopeia*. Discovering on her arrival that my father (*right*) had been called up, she accompanied him (illegally) to Kenya in the station wagon Lucy Lockett, seen here (*far right*) on a makeshift bridge where my mother is washing her face in the river, and at breakfast-time in one of their many camps (*below*).

One of my father's training locations coincided with Baden-Powell's funeral and he was invited, as a former Scout, to be a pall-bearer. I think he looks very dashing in his KAR uniform, marching next to Lord Erroll (out of step) who was murdered soon after.

To signal landmarks in family life, my mother had the custom of painting big tableaux representing scenes and events. This is a small part of one called 'The Ways that We Went', which she did for her Golden Wedding in 1989. Alongside generic African scenes are my father's armoured car in Somaliland, my mother and me striding into my life together, a sandy Lake Nyasa beach, Hookariah my pet chameleon, Percy our pet bushbaby, and our house at Makwapala with me pushing Sarah in the lorry towards Tui the dachshund.

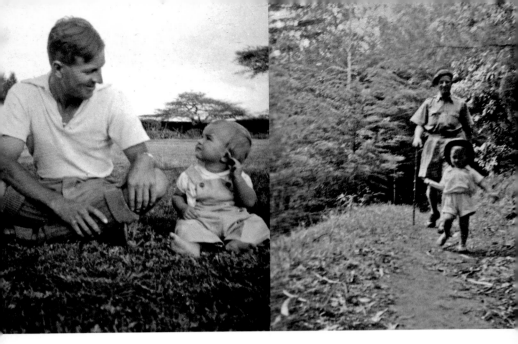

I evidently looked up to my father from an early age, and accompanied him in climbing the lower slopes of Kilimanjaro. Baraza kindly tolerated my dogged help in pushing my pram. Later, we moved to Makwapala in Nyasaland (*lower right*) where I seem to have grown bored with the sewing class my mother was conducting in the garden. In 1946, on a brief leave, we stayed with my grandparents in England. During this time my uncle Bill and aunt Diana (middle row left, next to my parents) married at Mullion, and the whole family picnicked at Kynance Cove.

On our return to Nyasaland we lived in Lilongwe, where my parents bought Creeping Jenny, our first new car. I was sent to board at the Eagle School in Southern Rhodesia. In the picture here, Tank (the headmaster) is in the centre with Coppers (matron) and Dick (another teacher) on his right. I'm the very small boy third from left in the same row and David Glynn, also small, is in the mirror position the other side, next to Wattie who is next to Paul. David and I collected the beautiful swallowtail butterflies which he mysteriously called 'Daddy Xmas'.

claim, but no single Oxford college comes close to Trinity Cambridge, not even Balliol, which tops the table of Nobel Prizes for Oxford colleges. My father, I have just realized, is one of few people to have studied at both Balliol, Oxford, and Trinity, Cambridge.

At both Oxford and Cambridge, the relationship between the colleges and the university bears the same uneasy tension as that between the federal and state governments in the USA. The rise of science has increased the power and importance of the 'federal government' (university), because science is too big an enterprise to be handled by each college separately (though one or two of them tried to go it alone in the nineteenth century). The science departments belong to the university, and it was the Zoology Department rather than the college that was to dominate my life at Oxford.

That porter must have been one of the first people to call me 'Mr Dawkins' (let alone 'Sir') – treat me as an adult – and I wasn't used to it. I think it was characteristic of my generation of undergraduates that we worked rather self-consciously at appearing to be more adult than we were. Later generations of undergraduates have tended towards the opposite, dressing scruffily with hoods or baseball caps, loosely slung rucksacks and sometimes even more loosely slung jeans. But my generation favoured tweed jackets with leather elbow patches, smart waistcoats, corduroy trousers, trilby hats, moustaches, ties, even bow ties. Some (not I, despite the example of my father) put the finishing touches to this image by smoking a pipe. These affectations may have been prompted by the fact that many of my fellow freshmen really were two years older; for my cohort was almost the first of the post-war generations not to be called up for military service. Those of us who came straight from school in 1959 were boys, sharing lectures, quadrangles and a dining hall with militarily trained *men*, and this perhaps raised our aspiration to grow up and be taken seriously as adults. We left Elvis behind and listened to Bach or the Modern Jazz Quartet. We solemnly intoned Keats and Auden and Marvell to each other. Chiang Yee captured the mood in his charming book from a slightly earlier era, *The Silent Traveller in Oxford*,[1] when he drew, in his elegant Chinese style, a pair of freshmen bounding, two steps at a time, up their college staircase. His deliciously perceptive caption read: 'I could tell that they were freshmen because I heard one say to the other, "Do you read much Shelley?"'

The claim that army service turns boys into men is the basis of a lovely story about Maurice Bowra, legendary Warden of Wadham College (anecdotes about Bowra are so numerous as to be best avoided, but this is an especially charming one). Immediately after the war, he was interviewing a young man for a place at the college.

'Sir, I have been away at the war, and I have to confess that I have forgotten all my Latin. I cannot pass the Latin exam to qualify for entrance.'

'Oh, don't you worry about that, dear boy, war counts as Latin, war counts as Latin.'

My older colleagues back from National Service in 1959 were not literally 'battle-hardened' like Bowra's entrance candidate, but they had an unmistakable air of being worldly-wise and grown up, in a way that I was not. As I said, I think that those of my generation who affected pipe-smoking, bow ties and neatly trimmed moustaches may have been struggling to keep up with the military veterans. Am I right in suspecting that today's undergraduates aspire in the opposite direction, towards juvenilization? On the first day of a new university year, a modern college noticeboard is likely to have notices saying things like this: 'Freshers! Feeling lonely? Lost? Missing Mum? Do drop in for coffee and a chat. We love you.' Such cosseting invitations would have been inconceivable on the noticeboard of my first term, which was more likely to carry

---

[1] Chiang Yee, *The Silent Traveller in Oxford* (London, Methuen, 1944).

announcements calculated to make me feel I had arrived in the adult world: 'Would the "gentleman" who "borrowed" my umbrella . . .'

I had applied to read biochemistry. The tutor who interviewed me, the kindly Sandy Ogston, who later became Master of Trinity, declined – thank goodness – to let me in as a biochemist (perhaps because he was one himself and would have had to teach me) but offered me a place to read zoology instead. I accepted gratefully, and it turned out to be the perfect course for me. Biochemistry could not have captured my enthusiastic interest the way zoology did: Dr Ogston was as wise as his venerable grey beard suggested.

Balliol had no tutorial fellow in zoology, so I was sent out of college to the wonderfully convivial Peter Brunet in the Department of Zoology. He would be responsible for tutoring me or for arranging tutorials with others. One incident in an early tutorial with Dr Brunet may have marked the beginning of my weaning away from a school attitude to learning in favour of a university one. I asked Dr Brunet a question about embryology. 'I don't know,' he mused, sucking on his pipe. 'Interesting question. I'll ask Fischberg and report back.' Dr Fischberg was the department's senior embryologist, so this was an entirely reasonable response. At the time, however, I was so impressed by Dr Brunet's attitude that I wrote to my parents about it. My tutor didn't know the answer to a question and was going to ask an expert colleague and report back to me! I felt that I'd joined the big boys.

Michael Fischberg was from Switzerland, with a very strong Swiss German accent. His lectures made frequent mention of things called 'tonk bars' and I think most of us wrote 'tonk bars' in our lecture notes before we finally saw the phrase written down: 'tongue bars', a feature of embryos at a certain stage of development. Endearingly, while at Oxford Dr Fischberg developed a great enthusiasm for our English national game, founding and captaining the departmental team. He had a most unusual bowling action.

Unlike a baseball pitcher, a cricket bowler has to keep his arm straight. Throwing is strictly forbidden: you must not bend your arm. Given this constraint, the only way to propel the ball at any speed is to run and then bowl while still running. The fastest bowlers in the world, such as the terrifying Jeff Thomson ('Tommo') of Australia, have achieved ball velocities of 100 mph (comparable to a baseball pitcher with his bent arm), and they do it by running very fast before releasing the ball with a straight overarm action in graceful rhythm with their running. Not Dr Fischberg. He stood rigidly to attention facing the batsman, raised his straight bowling arm horizontally to take careful aim at the wicket, then swung it over in a single arc and let go of the ball at the top.

I was hopelessly bad at cricket, but was sometimes cajoled into playing for Zoology when they couldn't find anybody better and were really desperate. I do, though, quite enjoy watching cricket, fascinated by the strategy of a captain placing his fielders around the batsman – like a chess master deploying his pieces to encircle the king. The best cricketer I ever saw playing in the Oxford University Parks was the Nawab of Pataudi ('Tiger'), the Oxford captain and my exact contemporary at Balliol. As a batsman, the effortless way he steered the ball to outwit the fielders was sublime. But it was as a fielder himself that he especially impressed me. On one occasion a batsman hit the ball and called for what must have seemed like an easy run. Then he noticed that the fielder charging down upon the ball was Tiger Pataudi, and he frantically shrieked to his partner to go back to his crease. Sadly, Tiger later lost one eye in a car accident and had to change his stance to bat monocularly, but he was still good enough to captain India.

I said that Oxford was the making of me, but really it was the tutorial system, which happens to be characteristic of Oxford and Cambridge. The Oxford zoology course also had lectures and laboratory classes, of course, but these were no more remarkable than those at any other university. Some lectures were good, some were bad, but it scarcely made any difference to me because I hadn't yet worked out the point of going to a lecture. It is not to imbibe information, and there is therefore no point in doing what I did (and what virtually all undergraduates do), which was take notes so slavishly that there was no attention left over for thinking. The only time I departed from this habit was once when I had forgotten to bring a pen. I was much too shy to borrow a pen from the girl sitting next to me (having been to a single-sex school, and shy to boot, I was in boyish awe of all girls at the time, and if I was too timorous to borrow a pen you can imagine how often I dared approach them for anything more interesting than that). So, for that one lecture I took no notes and just listened – and thought. It was not an unusually good lecture, but I got more out of it than from other lectures – some of them much better ones – because my lack of pen freed me to listen and think. But I didn't have the sense to learn my lesson and refrain from taking notes at subsequent lectures.

Theoretically the idea was to use your lecture notes in revision, but I never looked at mine ever again and I suspect that most of my colleagues didn't either. The purpose of a lecture should not be to impart information. There are books, libraries, nowadays the internet, for that. A lecture should inspire and provoke thought. You watch a good lecturer thinking aloud in front of you, reaching for a thought, sometimes grabbing it out of the air like the celebrated historian A. J. P. Taylor. A good lecturer thinking aloud, reflecting, musing, rephrasing for clarity, hesitating and then grasping, varying the pace, pausing for thought, can be a role model in how to think about a subject and how to transmit a passion for it. If a lecturer drones information as though reading it, the audience might as well read it – possibly in the lecturer's own book.

I exaggerate a little when I advise never to take notes. If a lecturer produces an original thought, something striking that makes you think, then by all means write yourself a memo to think again about it later, or look something up. But struggling to record a piece of every sentence the lecturer utters – which is what I tried to do – is pointless for the student and demoralizing for the lecturer. When lecturing to a student audience today, all I notice is a sea of tops of heads, bowed over notebooks. I prefer lay audiences, literary festivals, memorial lectures, guest lectures at universities where if the students come it is because they want to and not because it is on their syllabus. At such public lectures, the lecturer sees not bowed heads and scribbling hands but alert faces, smiling, registering comprehension – or the reverse. When lecturing in America, I get quite cross if I hear that some professor has *required* students to attend my lecture for 'credit'. I'm not keen on the idea of 'credit' at the best of times, and I actively hate the idea that students are getting credit for listening to me.

Niko Tinbergen, my later mentor, entered my life as the lecturer on molluscs. He announced no special affinity for that group save a fondness for oysters, but he played along with the department's tradition of handing out a phylum to each lecturer, more or less at random. From those lectures, I recall Niko's swift blackboard drawings; his deep voice (surprisingly deep for a small man), accented but not obviously Dutch; and his kindly smile (avuncular, as I thought it then, although he must have been much younger than I am now). In the following year he again lectured to us, this time on animal behaviour, and the avuncular smile broadened with enthusiasm for his own subject. In that heyday of his research group in the gull colony at Ravenglass in Cumberland, I was enchanted by his film on eggshell removal by black-headed gulls. I especially liked his method of plotting graphs – laying out tent-poles on the sand for axes, with strategically placed eggshells for data points. How very Niko. How very un-PowerPoint.

After each lecture there was a practical class in the laboratory. I had no aptitude for practical work, and – so young and immature was I – the opposite sex was even more of a distraction in the laboratory than in lectures. It was really only the tutorial system that educated me, and I shall forever be grateful to Oxford for this unique gift – unique because, at least where science subjects were concerned, I think even Cambridge was not equal in this respect. The Cambridge Natural Science Tripos Part I, which occupies the first two years of the undergraduate course, is commendably broad but in consequence it cannot give the student the exhilarating experience, as Oxford does, of becoming a world authority – I mean it only slightly short of literally – on a set of (admittedly very narrow) subjects. I explained this in an essay which was published in various places and definitively in a book called *The Oxford Tutorial: 'Thanks, you taught me how to think'*.[1] Parts of the following paragraphs are derived from this article.

I made the point there that our Oxford course was not 'lecture-driven' in the way that many undergraduates like their studies to be, feeling that they should be examined on, and only on, topics directly covered in lectures. On the contrary, when I was an undergraduate the entire subject of zoology was fair game for the examiners. The only constraint was an unwritten convention that the exam in any one year should not depart unfairly from the general precedent of previous years. And tutorials, too, were not 'lecture-driven' (as I fear they may be today); they were zoology-driven.

In my penultimate term Peter Brunet managed to secure for me the rare privilege of tutorials with Niko Tinbergen himself. Since he was solely responsible for all the lectures in animal behaviour, Dr Tinbergen would have been well placed to give 'lecture-driven'

---

[1] 'Evolution in biology tutoring?', in David Palfreyman, ed., *The Oxford Tutorial: 'Thanks, you taught me how to think'* (Oxford Centre for Higher Education Policy Studies, 2001; 2nd edn 2008). When the essay first appeared (in *The Oxford Magazine*, No. 112, Eighth Week, Michaelmas Term 1994), it bore the 'deliberately graceless' title 'Tutorial-Driven', in reflection of the 'lecture-driven' teaching I was criticizing.

tutorials. I need hardly say that he didn't. Each week my tutorial assignment was to read one DPhil (Oxford-speak for PhD) thesis. My essay was to be a combination of DPhil examiner's report, review of the history of the subject in which the thesis fell, proposal for follow-up research, and theoretical and philosophical discussion of the issues that the thesis raised. Never for one moment did it occur to either tutor or pupil to wonder whether this assignment would be directly useful for answering some exam question.

Another term Peter Brunet, recognizing that my bias in biology was more philosophical than his own, arranged for me to have tutorials with Arthur Cain, effervescently brilliant rising star of the department, who went on to become Professor of Zoology at Manchester and later Liverpool. Far from these tutorials being driven by any lectures on our course, Dr Cain had me reading nothing but books on history and philosophy. It was up to me to work out the connections between zoology and the books that I was reading. I did, and I loved it. I'm not saying that my juvenile essays on the philosophy of biology were any good – with hindsight I know they weren't – but I can say that I have never forgotten the exhilaration of writing them, or the feeling of being a real scholar as I read in the library.

The same is true of my more mainline essays on standard zoological topics. I have no memory of whether we had a lecture on the water-vascular system of starfish. Probably we did, but that fact had no bearing upon my tutor's decision to assign an essay on the topic. The starfish water-vascular system is one of many highly specialized topics in zoology that I now recall for the same reason – that I once wrote an essay on them. Starfish don't have red blood; instead, they have piped sea water, constantly circulated through an intricately plumbed system of tubes which form a ring around the centre of the star and lead off in branches down each of the five arms. The piped sea water is used in a unique hydraulic pressure system, operating the many hundreds of tiny tube feet arrayed along the five arms. Each tube foot ends in a little gripping sucker, and these shuttle back and forth in collusion to pull the starfish along in a particular direction. The tube feet don't move in unison but are semi-autonomous and, if the circum-oral nerve ring that gives them their orders should chance to become severed, the tube feet in different arms can pull in opposite directions and tear the starfish in half.

I remember the bare facts about starfish plumbing, but it is not the facts that matter. What matters is the way in which we were encouraged to discover them. We didn't just mug up a textbook: we went into the library and looked up books old and new; we followed trails of original research papers until we had made ourselves as nearly world authorities on the topic as it is possible to become in one week (nowadays one would do much of this work on the internet). The encouragement provided by the weekly tutorial meant that one didn't just *read* about starfish hydraulics, or whatever the topic was: for that one week, I remember that I slept, ate and dreamed starfish hydraulics. Tube feet marched behind my eyelids, hydraulic *pedicellariae* quested and sea water pulsed through my dozing brain. Writing my essay was the catharsis, and the tutorial was the justification for the entire week. And then the next week there would be a new topic and a new feast of images to be conjured up in the library. We were being educated . . . And I believe it is largely to this week-by-week training that I owe such writing ability as I may be judged to possess.

The tutor for whom I wrote the starfish essay was David Nichols, who went on to become Professor of Zoology at Exeter. Another notable tutor who shaped me as a young zoologist was John Currey, later Professor of Zoology at York University. He introduced me to, among other things, his – and now my – favourite example of revealingly bad 'design' in animals: the recurrent laryngeal nerve. As

I explained in *The Greatest Show on Earth,* instead of going directly from the brain to its end organ the larynx, this nerve makes a detour (in the case of the giraffe, a spectacularly long detour) down into the chest, where it loops around a large artery before proceeding back up the neck to the larynx. This is eloquent of terribly bad design, but is completely explicable the moment you forget design and start thinking in terms of evolutionary history instead. In our fishy ancestors the shortest route for the nerve was posterior to the then equivalent of that artery, which in those early days supplied one of the gills. Fish don't have necks. When necks started to lengthen on land, the artery gradually moved backwards relative to the head, step by tiny step through evolutionary time further away from brain and larynx. The nerve kept abreast – kind of literally – making at first only a small detour but then, as evolution progressed, a longer and longer detour until, in a modern giraffe, its diverted route is a matter of several metres. Just a few years ago, as part of a television documentary, I was privileged to assist in a dissection of this remarkable nerve in a giraffe that had unfortunately died a few days earlier.

My genetics tutor was Robert Creed, pupil of the eccentric and misogynistic aesthete E. B. Ford, himself heavily influenced by the great R. A. Fisher, whom we were all taught by Ford to revere. I learned from those tutorials, and from Dr Ford's own lectures, that genes are not atomistically separate from each other, where their effects on bodies are concerned. Rather, a gene's effect is conditioned by the 'background' of the other genes in the genome. Genes modify one another's effects. Later, when I became a tutor myself, I devised an analogy to try to explain this to my pupils. The body is represented by the shape of a bed-sheet, hanging approximately horizontally by thousands of strings attached to an array of hooks in the ceiling. Each string represents one gene. A mutation in the gene is represented by a change in the tension in that string's attachment to the ceiling. But – here is the important part of the analogy – each string is not isolated in its attachment to the sheet hanging below it. Rather, it is tangled up with lots of other strings, in a complicated cat's cradle. This means that when a mutation occurs in any one 'gene' (change of tension in its attachment to the ceiling hook), the tensions in all the other strings with which it is entangled change at the same time, in a series of knock-on effects throughout the cat's cradle. And the shape of the sheet (the body) is consequently influenced by the interaction of all the genes, not by each gene working separately on its 'own' little part of the sheet. In fact, no gene does 'own' any single part of the sheet. The body is not like a butcher's diagram, with 'cuts' of the body corresponding to particular genes. Rather, a gene may influence the whole body in interaction with other genes. An elaboration of the parable introduces environmental – non-genetic – influences tugging on the cat's cradle from the side.

From Arthur Cain, whom I mentioned above, I learned to dissent from the still fashionable trashing of numerical systems for classifying animals by mathematical measurement of the similarities and differences between them. Quite separately, I also learned from Dr Cain to be impressed by the power of natural selection to produce adaptations of extreme perfection – notwithstanding important and interesting exceptions such as the recurrent laryngeal nerve, just mentioned. Both these lessons set me somewhat at odds with certain orthodoxies, which still dominate the world of zoology. Arthur also taught me to be sparing in my use of the word 'mere' – an exercise in consciousness-raising that has stayed with me ever since. 'Humans are not *mere* bags of chemicals . . .' Well, of course they are not, but when you have said that you have said nothing interesting, and the word 'mere' is supererogatory. 'Humans are not *mere* animals . . .' What have you just said that is more than trite? What weight does the word 'mere' carry in that sentence? What is 'mere' about an animal? You haven't said anything meaningful. If you intend to mean something, say it.

Arthur also told me a never-forgotten story about Galileo, which summarizes what was new about Renaissance science. Galileo was showing a learned man an astronomical phenomenon through his telescope. This gentleman said, approximately: 'Sir, your demonstration with your telescope is so convincing that, were it not that Aristotle positively states the contrary, I would believe you.' Today it amazes us – or ought to – that anybody could possibly reject real observational or experimental evidence in favour of what some supposed authority had simply asserted. But that's the point. That is what has changed.

For us zoologists, unlike undergraduates reading history or English or law, tutorials almost never happened in our college, or indeed in any college. Nearly all were in the Department of Zoology, a rambling up-stairs and down-dale appendage to the University Museum. It was this warren of rooms and corridors which, as I have already mentioned, was the centre of my being. This was very different from the typical experience of an Oxford undergraduate reading a non-scientific subject, for whom the college was the centre of existence. Old-style college tutors think that tutoring outside the college walls is a sort of second-best. My experience suggests exactly the opposite. It was refreshing to have a different tutor every term, for reasons that seem to me almost too obvious to specify.

I did have friends in Balliol, most of whom were reading non-scientific subjects. Nicholas Tyacke (with whom I later shared lodgings, and who became a professor of history at University College, London) and Alan Ryan (who became a distinguished political philosopher and Warden of New College) were on my staircase. As it happened, several of my friends were in the college's acting fraternity, which led me to see some amateur dramatic productions. One of the most moving theatrical evenings I ever experienced was a Balliol College Dramatic Society production of Robert Ardrey's *Shadow of Heroes*, about the Hungarian revolution of 1956. More light-hearted were the Balliol Players, a travelling company who each year would put on a pastiche production of an Aristophanes play. I think that when they started in the 1920s the Players did Aristophanes straight, even in Greek. But the tradition changed, and by my time they were rewriting Aristophanes into revues satirizing modern politics. The leading lights of the Players in my time were Peter Snow, who became a familiar face on television, and John Albery, a witty and talented member of the famous theatrical dynasty, who later became Master of University College, Oxford. John Albery did a splendid General Montgomery ('Now God said – and I agwee with him . . .'), and Peter Snow an equally memorable General de Gaulle: '*La gloire . . . la victoire . . . l'histoire . . . et . . . la plume . . . de ma tante.*' Jeremy Gould scarcely had to act at all to do Harold Macmillan singing 'My birthday honours list is certain to contain . . . And plenty of OBEs . . .' It was the time of the twilight of empire, and the Players did a lovely valedictory song, presumably written by John Albery, of which I remember only five lines:

> Sunset and the evening star
> From Aden to Zanzibar.
> The bonds of the Empire sundering
> And final salutes are thundering
> And man will not cease his wondering . . .

The same theatrical set introduced me to the Victorian Society, in whose company I spent some of my happiest times in Balliol. We met once or twice a term to sing music-hall songs to piano accompaniment, while sipping port. A master of ceremonies would call up soloists one by one to sing their special songs, and we'd all join in the chorus. Mostly they were cheerful, cheeky songs ('Where did you get that hat?' 'Don't have any more, Mrs Moore'; 'You can't do that there 'ere'; 'I'm 'Enery the Eighth I am'; 'My old man said follow the van') interspersed with some sentimental weepies, for which tissues would be handed out ('She's only a bird in a gilded cage'; 'Silver threads among the gold'), and the evening would end with jingoistic patriotism ('Soldiers of the Queen'; 'We don't want to fight, but by jingo if we do . . . The Russkies shall not have Constantinople'). If there's one experience from Balliol days that I would dearly love to relive, it would be an evening with the Victorian Society.

It was much later in my life, but the nearest approach to such a reliving took place at the regular Friday evening sing-song at the Killingworth Castle pub in Wootton, a village just outside Oxford, to which I was introduced by my second wife Eve, mother of my beloved daughter Juliet. The music was British 'folk', not music hall, and the drink was beer, not port, but here I relived something of the atmosphere of the Victorian Society: a warm conviviality fuelled by music and community, more than by drink. The soloists and instrumentalists (guitar, squeezebox, penny whistle) on these Friday nights rotated between four or five regular performers or groups, all of them good in their different ways, all with their particular repertoires of songs, which were known to the regular chorus including Eve and me. For some songs quite stylish canons and descants would be produced, and – as with the Victorian Society – the chorus was always disciplined and up to a brisk tempo, very different from the usual 'Just a song at twilight' drunken dirge. We knew the more prominent members by private nicknames given them by Eve: 'Two Pints' (a large, bearded young man with a huge bass voice as muscular as the arms that raised his pints and took the collection for the musicians); 'Big Daddy' (a grandfatherly figure with an agreeable tenor, who sometimes volunteered 'Cock Robin' as a solo after the main soloists had finished); 'Maynard Smith' (a cheery, bespectacled fellow, named for his facial resemblance to the great scientist); 'the Incredible Hulk' (one of few who sang out of tune) and others.

Back in undergraduate days, my Balliol friends and I often went to the cinema, usually to the Scala in Walton Street: intellectual films by Ingmar Bergman, or Jean Cocteau, or Andrzej Wajda or other continental directors. I was especially affected by Ingmar Bergman's dark monochrome images in *Wild Strawberries* and *The Seventh Seal*, and the lyrical love scenes of *Summer Interlude* before it turned tragic. Films of that kind, and poetry to which my father introduced me – Rupert Brooke, A. E. Housman and above all the early W. B. Yeats – turned my young self into unrealistic, indeed deluded, byways of romantic fantasy. Like many a naive nineteen-year-old I fell in love – not with any particular girl, but with the idea of being in love. Well, there was a girl, and she happened to be Swedish, which chimed with my Bergman-led fantasies, but it was the idea of love itself, with me in the role of a tragic Romeo, that I loved. I moped over her for a ludicrously long time after she had returned to Sweden and – no doubt – had long forgotten her brief Summer Interlude with me.

I didn't finally lose my virginity until much later, at the rather advanced age of twenty-two, to a sweet cellist in London, who removed her skirt in order to play to me in her bedsitter (you can't play the cello in a tight skirt) – and then removed everything else. It is fashionable to decry one's first such experience but I shall not. It was wonderful, and what I chiefly remember is the feeling of atavistic fulfilment: 'Yes, of course, *this* is what it was always going to feel like. This is the way it was going to be from the beginning of time.' It isn't difficult for a biologist to explain why nervous systems evolved in such a way as to make sexual congress one of the consistently greatest experiences life has to offer. But explaining it doesn't make it any the less wonderful – just as Newton's spectral unweaving never diminishes the glory of the rainbow. And it doesn't matter how many rainbows you see throughout your life. The glory is reinvented afresh, and the heart leaps up every time. But I'll say no more on the subject, and will betray no confidences. It isn't that kind of autobiography.

Wordsworth, as it happened, was never a favourite of mine, but I would like to quote here a few fragments of some of the poems that did move me as a young man. These verses were an important part of making me what I am, and they were all (in some cases still are) word-perfect in my memory.

> Breathless, we flung us on the windy hill,
> Laughed in the sun, and kissed the lovely grass.
> You said, 'Through glory and ecstasy we pass;
> Wind, sun, and earth remain, the birds sing still,
> When we are old, are old . . .' 'And when we die
> All's over that is ours; and life burns on
> Through other lovers, other lips,' said I,
> 'Heart of my heart, our heaven is now, is won!'
> 'We are Earth's best, that learnt her lesson here.
> Life is our cry. We have kept the faith!' we said;
> 'We shall go down with unreluctant tread
> Rose-crowned into the darkness!' . . . Proud we were,
> And laughed, that had such brave true things to say.
> –And then you suddenly cried, and turned away.
>
> Rupert Brooke

> Tell me not here, it needs not saying,
> What tune the enchantress plays
> In aftermaths of soft September
> Or under blanching mays,
> For she and I were long acquainted
> And I knew all her ways.
>
> A. E. Housman

I dreamed that I stood in a valley, and amid sighs,
For happy lovers passed two by two where I stood;
And I dreamed my lost love came stealthily out of the wood
With her cloud-pale eyelids falling on dream-dimmed eyes:
I cried in my dream, *O women, bid the young men lay*
Their heads on your knees, and drown their eyes with your hair,
Or remembering hers they will find no other face fair
Till all the valleys of the world have been withered away.

W. B. Yeats

Heart handfast in heart as they stood, 'Look thither,'
Did he whisper? 'look forth from the flowers to the sea;
For the foam-flowers endure when the rose-blossoms wither,
And men that love lightly may die–but we?'
And the same wind sang and the same waves whitened,
And or ever the garden's last petals were shed,
In the lips that had whispered, the eyes that had lightened,
Love was dead.

A. C. Swinburne

My father kept a loose-leaf folder in which he bound a large number of his favourite poems, all copied out in his own hand. My own taste in poetry was strongly influenced by this private anthology, which my mother still possesses. I was touched to learn that it originated in letters to her in their early twenties, sent from Cambridge where he was doing postgraduate studies, each poem enclosed with a letter and preserved by her.

But, to my own undergraduate days and my thoughts of what should come next: I don't think I ever seriously contemplated joining my father in farming. Increasingly, I wanted to stay on at Oxford to do a research degree. I didn't have any very clear idea of what might follow after that, or of what kind of research I wanted to do. Peter Brunet offered me a biochemical project, and I gratefully signed up to it and studied the relevant research literature, though without much enthusiasm. But then I went for tutorials with Niko Tinbergen on animal behaviour – and my life changed. Here was a subject I could really think about: a subject with philosophical implications. Niko was apparently impressed by me: his end-of-term report to my college said I was the best undergraduate he had ever tutored – although that verdict must be tempered by the fact that he didn't do much undergraduate tutoring. Anyway, it raised my courage to the point of asking him whether he would like to take me on as his research student, and to my enduring delight he said yes. My future was assured, for the next three years at least. And for the rest of my life, now that I think about it.

LEARNING
THE TRADE

PERHAPS all scientists recall their graduate student years as an idyll. But surely some research environments are more idyllic than others, and I think there was something special about the Tinbergen group at Oxford in the early 1960s. Hans Kruuk has captured the atmosphere in his affectionate but not hagiographic biography, *Niko's Nature*.[1] He and I arrived too late for the heroic 'hard core' period described by Desmond Morris, Aubrey Manning and others, but I think our time resembled it – though we saw less of Niko himself, because his room was in the main Zoology Department while all the rest of us were housed in the annexe at 13 Bevington Road, a tall, narrow house in north Oxford, about half a mile from the main Zoology building tacked onto the University Museum in Parks Road.

The senior figure in 13 Bevington Road was Mike Cullen, probably the most important mentor in my life – and I believe most of my contemporaries in the Animal Behaviour Research Group (ABRG) would agree. To try to explain the debt that all of us owe to this remarkable man, I can do no better than quote the closing

[1] Hans Kruuk, *Niko's Nature: The Life of Niko Tinbergen and his Science of Animal Behaviour* (Oxford, Oxford University Press, 2003).

words of the eulogy that I spoke at his memorial service in Wadham
College, Oxford, in 2001.

He did not publish many papers himself, yet he worked
prodigiously hard, both in teaching and research. He was
probably the most sought-after tutor in the entire Zoology
Department. The rest of his time – he was always in a hurry and
worked a hugely long day – was devoted to research. But seldom
his own research. Everybody who knew him has the same story to
tell. All the obituaries told it, in revealingly similar terms.

You would have a problem with your research. You knew
exactly where to go for help, and there he would be for you. I see
the scene as yesterday. The lunchtime conversation in the crowded
little kitchen at Bevington Road, the wiry, boyish figure in the red
sweater, slightly hunched like a spring wound up with intense
intellectual energy, sometimes rocking back and forth with concentration. The deeply intelligent eyes, understanding what you
meant even before the words came out. The back of the envelope
to aid explanation, the occasionally sceptical, quizzical tilt of the
eyebrows, under the untidy hair. Then he would have to rush off
– he always rushed everywhere – perhaps for a tutorial, and he
would seize his biscuit tin by its wire handles, and disappear. But
next morning the answer to your problem would arrive, in Mike's
small, distinctive handwriting, two pages, often some algebra,
diagrams, a key reference to the literature, sometimes an apt verse
of his own composition, or a fragment of Latin or classical Greek.
Always encouragement.

We were grateful, but not grateful enough. If we had thought
about it we would have realized, he must have been working on
that mathematical model of my research all evening. And it isn't
only me for whom he does this. Everybody in Bevington Road
gets the same treatment. And not just his own students. I was
officially Niko's student, not Mike's. Mike took me on, without
payment and without official recognition, when my research

became more mathematical than Niko could handle. When the time came for me to write my thesis, it was Mike Cullen who read it, criticized it, helped me polish every line. And all this, while he was doing the same thing for his own official students.

When (we all should have wondered) does he get time for ordinary family life? When does he get time for his own research? No wonder he so seldom published anything. No wonder he never wrote his long-awaited book on animal communication. In truth, he should have been joint author of just about every one of the hundreds of papers that came out of 13 Bevington Road during that golden period. In fact, his name appears on virtually none of them – except in the Acknowledgements section . . .

The worldly success of scientists is judged – for promotion or honours – by their published papers. Mike did not rate highly on this index. But if he had consented to add his name to his students' publications, as readily as modern supervisors insist on putting their names on papers to which they contribute much less, Mike would have been a conventionally successful scientist, lauded with conventional honours. As it is, he was a brilliantly successful scientist in a far deeper and truer sense. And I think we know which kind of scientist we really admire.

Oxford sadly lost him to Australia. Years later, in Melbourne, at a party for me as visiting lecturer, I was standing, probably rather stiffly, with a drink in my hand. Suddenly, a familiar figure shot into the room, in a hurry as ever. The rest of us were in suits, but not this familiar figure. The years vanished away. Everything was the same – though he must have been well into his sixties, he seemed still to be in his thirties – the glow of boyish enthusiasm, even the red sweater. Next day he drove me to the coast to see his beloved penguins, stopping on the way to look at giant Australian earthworms, many feet long. We tired the sun with talking – not, I think, about old times and old friends, and certainly not about ambition, grant-getting and papers in *Nature*, but about new science and new ideas. It was a perfect day, the last day I saw him.

We may know other scientists as intelligent as Mike Cullen – though not many. We may know other scientists who were as generous in support – though vanishingly few. But I declare, we have known nobody who had so much to give, combined with so much generosity in giving it.

I almost wept when I spoke that eulogy in Wadham chapel, and I almost wept again just now when rereading it twelve years later.

I don't know whether the camaraderie of 13 Bevington Road was exceptional, or whether all groups of graduate students nurture a similar *esprit de corps*. I suspect that being housed in a separate annexe rather than in a large university building improves the social dynamics. When the ABRG (and other outliers such as David Lack's Edward Grey Institute of Field Ornithology and Charles Elton's Bureau of Animal Populations) eventually moved into the present concrete monster on South Parks Road, something, I believe, was lost. But it may just be that by then I was older and more weighed down by responsibilities. Whatever the reason, I retain a loyal affection for 13 Bevington Road and my comrades of those times who foregathered at the Friday evening seminars, or in the lunch room, or over the bar billiards table in the Rose and Crown: Robert Mash, whose epidemic sense of humour I later recalled in my foreword to his book *How to Keep Dinosaurs*;[1] Dick Brown, chain smoking, hard drinking and implausibly rumoured to be religious; Juan Delius, whose deliriously eccentric brilliance never ceased to amuse; Juan's supernormally delightful wife Uta who gave me German lessons; the tall, blond Dutchman Hans Kruuk, who later wrote Niko's biography; the Scotsman Ian Patterson; Bryan Nelson the gannet man, known to me in my first six months only from the enigmatic notice on his door, 'Nelson is on the Bass Rock'; bearded Cliff Henty; David McFarland, Niko's eventual successor who,

---

[1] Robert Mash, *How to Keep Dinosaurs* (London, Orion, 2005).

although based in the Psychology Department, was a sort of honorary member of our group because his vivacious wife Jill was Juan's research assistant, and the couple had lunch in Bevington Road every day; Vivienne Benzie, who introduced the sunny New Zealand girls Lyn McKechie and Ann Jamieson as yet other honorary members of the lunch group; Lou Gurr, another smiling New Zealander; Robin Liley; the jovial naturalist Michael Robinson; Michael Hansell, who later shared a flat with me; Monica Impekoven, with whom I wrote a paper later; Marian Stamp, whom I was to marry; Heather McLannahan, Robert Martin, Ken Wilz; Michael Norton-Griffiths and Harvey Croze, who later formed a consulting partnership in Kenya; John Krebs, who later collaborated with me in writing three papers; the daredevil Iain Douglas-Hamilton, unwilling exile from Africa while he wrote his thesis on elephants; Jamie Smith, with whom I wrote a paper on optimal foraging in tits; Tim Halliday the newt man, Sean Neill with his lovingly restored Lagonda and gift for drawing cartoons, Lary Shaffer, master photographer, and other friends whom I apologize for omitting.

The Friday evening seminars were the highlight of the week for the Tinbergen group. They lasted two hours and frequently spilled over into the following Friday, but the time flashed by because, instead of the soporific formula of an hour spent listening to one speaker's voice followed by questions at the end, our two hours were enlivened by argument throughout. Niko set the tone by interrupting almost before the speaker could complete his first sentence: '*Ja, ja*, but what do you mean by . . .?' This wasn't as irritating as it sounds, because Niko's interventions always aimed at clarification and it was usually necessary. Mike Cullen's questions were more penetrating, better informed and more feared. Other notable contributors – brilliant in their idiosyncratic ways – were Juan Delius and David McFarland, but the rest of us chipped in without inhibition too, almost from the first day we were there. Niko encouraged that. He insisted on absolute clarity about the question we were asking in our research. I recall how shocked I was, on visiting our sister research group at Madingley in Cambridge, to hear one of the graduate students beginning to describe his research with the words: 'What I do is . . .' I had to restrain myself from imitating Niko's voice: '*Ja, ja*, but what is your *question*? Years later, I related this story when I gave a research seminar at Madingley. I refused to identify the culprit to a mock-scandalized Robert Hinde, the formidably intelligent and charismatic leader of the Madingley group who later became Master of St John's College, Cambridge, and my lips are sealed to this day.

The question Niko set for me was a version of the question often labelled with the 'nature or nurture?' cliché derived from *The Tempest*:

> A devil, a born devil, on whose nature
> Nurture can never stick . . .

Philosophers down the centuries have pondered the question. How much of what we know is natively built in, and to what extent is the young mind a blank slate, waiting to be written over, as John Locke believed?

Niko himself, like Konrad Lorenz (with whom he is credited with co-founding the science of ethology), was early associated with the 'nature' school of thought. His most famous book, *The Study of Instinct*,[1] which he later pretty much disowned, used 'instinct' as a synonym for 'innate behaviour', defined as 'behaviour that has not been changed by learning processes'. Ethology is the biological study of animal behaviour. Various schools of psychology also study animal behaviour, but with different emphasis. Psychologists

[1] N. Tinbergen, *The Study of Instinct* (Oxford, Clarendon Press, 1951).

historically tended to study animals like rats or pigeons or monkeys as substitutes for humans. Ethologists historically were interested in the animals in their own right, not as proxies for anything. Consequently they have always studied a much wider range of species, and they tend to emphasize the role of behaviour in the natural environment of the species. Ethologists also, as I have said, historically emphasized 'innate' behaviour, whereas psychologists were more interested in learning.

In the 1950s, a group of American psychologists started to take an interest in the works of the ethologists. Prominent among them was Daniel S. Lehrman, a big man with a deep knowledge of natural history as well as of psychology. He also spoke adequate German, which made him an effective bridge between the two approaches to animal behaviour.

In 1953 Lehrman wrote a very influential critique of the traditional ethological approach. He strongly criticized the whole notion of innate behaviour, not because he thought everything was learned (although some psychologists whom he quoted did), but because he thought it was in principle impossible to define innate behaviour: impossible to devise an experiment to demonstrate that any particular piece of behaviour is innate. Theoretically, the obvious method was the 'deprivation experiment'. Imagine if humans were given no verbal instruction in how to copulate and no opportunity to observe other species – not even the smallest inkling. Would they know how to do it when the opportunity finally presented itself? It's an intriguing question, and there might be telling anecdotes, perhaps about over-sheltered and naive Victorian couples. But in non-human animals we can do experiments. Deprivation experiments.

If you rear a young animal in deprived conditions without the opportunity for experience, and it still knows how to behave properly, that must mean the behaviour is innate, inborn, instinctive. Mustn't it? But Lehrman objected that you couldn't deprive the young animal of everything – light, food, air, etc. – and that it is never obvious how much deprivation is needed in order to satisfy the criterion of innateness.

The dispute between Lehrman and Lorenz got personal. Lehrman, whose family background was Jewish, caught Lorenz out in some suspiciously Nazi-inflected writings from the war years and did not shrink from mentioning this in his famous critique. Lorenz, on first meeting Lehrman after the critique was published, said (approximately): 'I thought from your writings that you must be a small, mean, wizened little man. But now that I see you are a BIG man [and Lehrman was indeed a very big man] we can be friends.' This avowal of friendship didn't stop Lorenz trying – Desmond Morris tells the story as an eye-witness from inside the car – to intimidate Lehrman by almost mowing him down with an enormous American car that he was driving in Paris.

But back to the controversy over nurture or nature. Male sedge warblers (to take just one example) have a complex and elaborate song, and they can perform it even when reared in isolation, never having heard another sedge warbler. The Lorenz–Tinbergen school would therefore have said it must be 'innate'. But Lehrman emphasized the complexity of developmental processes and always wondered whether learning was involved in some less obvious way. For Lehrman, it wasn't good enough to say that the young animal had been reared under deprived conditions. For him, the question was: 'Deprived of what?'

Since Lehrman's critique was published, ethologists have indeed discovered that many young songbirds, including sedge warblers, even when reared in isolation, *learn* to sing their correct species song by listening to their own fumbling efforts, repeating the good fumbles and discarding the bad. So that looks like nurture after all. But in that case, Lorenz and Tinbergen might reply, how do the young birds know which of their fumbles are good and which bad? Surely that 'knowledge' – a template for what their species song ought to sound like – has to be innate? All learning does is transfer the song pattern from the sensory part of the brain (the built-in template) to the motor side (the actual skill of singing the song).

Other species, by the way, such as the American white-crowned sparrow, also teach themselves to sing in this 'fumbling' way, but do need to have heard the species song earlier in life. It is as though the young bird takes a 'tape recording' before it can sing, and uses it as a template for teaching itself how to sing. And there are intermediates between the 'learned tape recording' and the 'innate tape recording' as templates for later learning.

This was the philosophical minefield into which Niko Tinbergen released me in 1962. I think he wanted to back away from his perceived association with Lorenz and saw me as a bridge towards the Lehrman camp. My experimental subject was to be not singing birds but baby chicks pecking. I did a series of experiments of which I'll mention only one here.

Baby chicks straight out of the egg start pecking at small objects, presumably looking for food. But how do they know what to peck at? How do they know what's good for them? One extreme would be for nature to endow them, before they have any experience at all, with a template picture of a grain of wheat in the brain. That's un-realistic, especially in an omnivore. Do wheat grains and mealworms and barleycorns and millet seeds and beetle larvae have anything in common, as opposed to boring and inedible marks and stains? Yes, they do. For one thing, they are solid.

How do you recognize something as solid? One way is by surface shading. Look at the photographs of moon craters overleaf. They are the same photograph, but one is rotated through 180° relative to the other. My guess is that on the left you will see hollow craters and on the right solid flat-topped hills – and the other way around if you swivel the book upside down. The illusion has been known for a long time. It depends upon a preconception about where the light is coming from: in effect, a preconception about the location of the sun. Solid objects tend to be brighter on the side nearest the sun, which will usually be approximately above. A photograph of a solid object can therefore look hollow if you turn it upside-down, and vice versa.

The sun is seldom *directly* overhead, but the general direction of its light is more likely to be down than up. Therefore any predator seeking solid objects as possible prey can use surface shading cues based upon that assumption. And on the other side of the predator–prey arms race, natural selection might well favour prey animals that manage to disguise their solidity by 'countershading'. Many species of fish are darker on top, lighter below, which tends to neutralize the natural tendency for sunlight to come from above, and thereby makes the fish look flatter. One fish, the 'upside-down catfish', is a genuine 'exception that proves the rule'. It habitually swims upside-down and, sure enough, it is *reverse countershaded*: darker on its belly than its back.

A Dutch student of Tinbergen called Leen De Ruiter did some

neat experiments on reverse countershaded caterpillars, who habitually rest upside-down. The upper picture on this page shows *Cerura vinula* in its normal position. It looks flat and inconspicuous. The lower picture shows what it looked like when De Ruiter turned its twig upside-down: much more conspicuous to my eyes and – more significantly – to the eyes of jays, when De Ruiter used them as experimental predators.

But none of this says anything about whether – in jays or

humans – the knowledge that the sun is normally overhead is innate or learned. The solid shading illusion seemed to me to provide a good opportunity to test the question, using baby chicks in deprivation experiments.

First, did chicks see the illusion? Apparently, yes. I photographed a half ping-pong ball lit asymmetrically, and printed the image to be about the size of a tempting grain or seed. When I viewed the photograph with the illuminated side at the top, the hemisphere looked solid. When I inverted the photograph, it didn't. When chicks were offered a choice between the two orientations, they strongly chose to peck at the apparently solid picture, the one lit from above. This suggested that chicks possess the same 'preconception' we do, that the sun is normally overhead.

So far, so good; but these chicks, though young, were not completely naive. They were three days old and had been feeding in normal overhead light during this time. They might have had time to learn the appearance of solid objects illuminated from above.

To test this I did a crucial experiment. I reared chicks with light coming from *below* and tested them under the same conditions. So, at the time of testing, they had never had any experience of overhead light. As far as they were concerned, the world into which they had hatched was a world with a sun underneath them. Every solid object they had ever seen, whether food objects or parts of other chicks, was lighter underneath than on top. I expected that, when tested with the two ping-pong ball photographs, they would prefer to peck at the one illuminated from below.

But I was delighted to be proved wrong. The chicks over-whelmingly pecked at the photograph illuminated from above. If you accept my interpretation, this means that the chicks are genetically equipped by ancestral natural selection with something equivalent to 'advance information': in the world in which they are to live, the sun will normally shine from above. My experiment had pinpointed a true example of innate information which is not reversed by a positive attempt to teach the contrary.

I can't think of any group of humans who habitually live with underfloor lighting. If they exist, it would be interesting to test them in the same way I tested my chicks. I thought about offering an intuitive guess as to what the result would be, but I honestly prefer not to place a bet. Wouldn't it be fascinating if we too saw the illusion innately? Having been surprised by the chicks, I'd be only slightly more surprised if humans did the same. We may never know, but there could be ways to do the experiment on very young babies. They don't peck, but they do fixate their eyes on objects that interest them, and you can measure that. Could a developmental psychologist offer babies a version of my ping-pong ball experiment and measure the time they spend staring at each of the two photographs? Would it be considered unethical to use underfloor lighting for a baby's room for the first few days of life? I can't see why, but who knows what the verdict of a modern 'ethical committee' might be?

In the end, my work on 'nature or nurture' constituted only a small part of my doctoral research,[1] and it was relegated to an appendix in my thesis. The main part of my thesis had little in common with it, except that it also involved pecking in chicks. And it was also an attempt to illustrate a point of philosophical interest – although taken from a different part of philosophy. It became possible through an improved technique for recording pecks.

Bevington Road, and especially its satellite research stations in the great gull colonies of the north, ran a system of 'slaves' – young unpaid volunteers who wanted a brief taste of the Tinbergen experience before going to university. Among them were Fritz

---

[1] R. Dawkins, 'The ontogeny of a pecking preference in domestic chicks', *Zeitschrift für Tierpsychologie*, 25 (1968), pp. 170–86.

Vollrath (who later returned to Oxford to head a flourishing group working on spider behaviour, and remains a close friend) and (also from Germany) Jan Adam. Jan and I found an immediate affinity, and we worked together. He had remarkable workshop skills – combining the very different virtues of my father and Major Campbell – and, fortunately, these were the days before health and safety regulations interfered to protect us from ourselves and sap our initiative. Jan and I had the freedom of the departmental workshops: lathes, milling machines, bandsaws and all. We (that is to say Jan, with me as willing apprentice – the younger brother syndrome again, I suppose) built an apparatus to automate the counting of chick pecks, using delicately hinged little pecking keys, elegantly made from scratch by Jan, with sensitive micro-switches. Previously, when working on the surface shading illusion, I had counted pecks by hand. Suddenly, I was in a position to collect huge quantities of data automatically. And this opened the door to a completely different kind of research, motivated by a different philosophy, Karl Popper's philosophy of science, which I learned from Peter Medawar.

As I have already explained, I had come to know of Medawar early on through my father, who was a schoolfriend of his. As British biology's star intellectual, Medawar came to give a visiting lecture at his old Oxford department when I was an undergraduate there, and I remember the excited buzz in the standing-room-only audience waiting for this tall, handsome, gracious figure to arrive ('This lecturer has never been thought ungracious in his life,' as a later critic said of him). The lecture prompted me to read Medawar's essays, later anthologized in *The Art of the Soluble* and *Pluto's Republic*,[1] and it was from them that I learned about Karl Popper.

I became intrigued by Popper's vision of science as a two-stage

---

[1] Peter Medawar, *The Art of the Soluble: Creativity and Originality in Science* (London, Methuen, 1967); *Pluto's Republic: Incorporating The Art of the Soluble and Induction and Intuition in Scientific Thought* (Oxford, Oxford University Press, 1982).

process: first the creative – almost artistic – dreaming up of a hypothesis or 'model', followed by attempts to *falsify* predictions deduced from it. I wanted to do a textbook Popperian study: dream up a hypothesis that might or might not be true, deduce precise mathematical predictions from it, and then try to falsify those predictions in the lab. It was important to me that the predictions should be mathematically precise. It was not enough to predict that a measurement X should be larger than Y. I wanted a model that would predict the exact value of X. And this kind of exact prediction demanded large quantities of data. Jan's apparatus for counting massive numbers of pecks gave me the opportunity. Instead of pecking at photographs of ping-pong balls, my birds pecked at little coloured hemispheres mounted on Jan's hinged windows, which triggered micro-switches. They preferred blue over red over green, but that wasn't what interested me. I wanted to know what governed each individual pecking decision, whichever colour it was directed towards. And this, of course, was only a specimen of a more general question about how decisions are made at any time by any animal.

Medawar elsewhere made the point that scientific research doesn't develop in the same orderly sequence as the final published 'story'. Real life is messier than that. In my own case it was so messy that I can't remember what gave me the idea for my 'Popperian' experiments. I remember only the finished story which, as Medawar would have expected, gives an implausibly tidy impression.

The finished story is that I dreamed up an imaginary 'model' of what might be going on inside a chick's head when it makes a decision between alternative targets, did some algebra to deduce precise, quantitative predictions from the model, then tested them in the lab. The model itself was a 'drive/threshold' model. I postulated that there was a variable ('drive' to peck) in the bird's head, whose graph was continuously wiggling up and down as the drive strengthened or weakened (perhaps at random; it b b b r b br b r r r b b r br b b b b b b b b b r b r r b b r br gr bg r b r g gr gb gb r g r b b bb

The wiggly line is the drive. When it is above only blue's threshold, all pecks are at blue. When it is above red's threshold, it is automatically above blue's too, and the two colours are choosen at random. When it is above green's threshold, all three colours are chosen at random because all three thresholds are exceeded.

didn't matter). Every time the drive happened to rise above the threshold for a colour, the bird was capable of pecking at that colour (something else, for which I developed and tested another model which I'll mention later, determined the *timing* of pecks). Blue, being a preferred colour, had a lower threshold than green. But if the drive rose above green's threshold, it automatically had to be above blue's threshold as well. What would the bird do then? I postulated that it would be indifferent between the two colours, since both thresholds were exceeded: it would 'toss a coin' to decide between them. So the model predicted that a bird's choices over a long period would consist of periods of pecking at only the preferred colour, interspersed with periods of choosing at random between the two. There would be no periods of consistently positive choice of the less preferred colour.

I didn't at first look directly at sequences of pecks. That was to come later, after I moved to California. I think the reason I didn't test sequences at first was as unambitious as the fact that Jan's apparatus could count pecks but not record the exact order in which they happened; and Jan himself had by now gone back to Germany, so wasn't there to modify his apparatus. I think, too, that I was simply seduced by the Popperian elegance of deducing a mathematical formula which would predict some measured quantity from some other measured quantities.

The chicks happened to prefer blue over red over green. I imagined an experiment in which I would present Blue versus Green, Blue versus Red, and Red versus Green, counting the proportion P of pecks to the preferred colour in each case. This would give me three numbers ($P_{BestWorst}$, $P_{BestMedium}$, $P_{MediumWorst}$). It's only to be expected that $P_{BestWorst}$ would be larger than either of the other two. But could the model predict precisely how much bigger? Could I deduce from the model a formula to predict exactly what $P_{BestWorst}$ should be, if I fed in $P_{BestMedium}$ and $P_{MediumWorst}$? Yes, that is exactly what I succeeded in doing. I defined algebraic symbols to stand for the time spent by the drive between various thresholds, did some school algebra (simultaneous equations as taught by Ernie Dow) to eliminate the unknown variables, and was pretty pleased when, at the end of pages of algebra, a simple, precise, quantitative prediction dropped out. The drive/threshold model predicts that

$$P_{BestWorst} = 2(P_{BestMedium} + P_{MediumWorst} - P_{BestMedium} \cdot P_{MediumWorst}) - 1.$$

I called this Prediction 1. The thing that interested me about Prediction 1 was that it is quantitatively precise.

So now to test it. Would the chicks obey the prediction? Yes: to my delight and amazement, in seven out of eight repeats of the experiment they did, very closely. The eighth experiment was way off, so much so that, to my acute embarrassment, when one of my papers was published in the journal *Animal Behaviour*,[1] the printer

---

[1] R. Dawkins, 'A threshold model of choice behaviour', *Animal Behaviour*, 17 (1969), pp. 120–33.

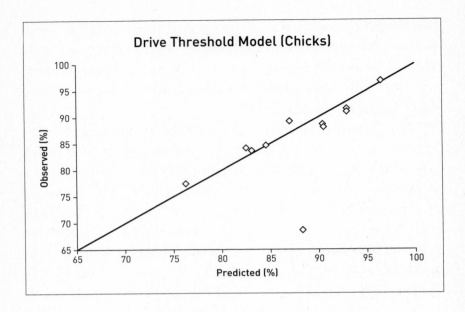

removed the relevant point from the graph, thinking it must be a blemish on the block! Fortunately the offending datum was clearly present in the accompanying table, otherwise I might have been accused of dishonesty. I did another set of experiments on chicks, involving not pecking but walking into chambers illuminated by light of different colours. The graph shown here combines the two sets of experiments and plots the observed against predicted percentages for all 11 chick experiments.

If the model's predictions were perfect, the points should all lie exactly along the diagonal line. With the exception of Experiment 8, as already mentioned, the Drive Threshold Model does a far better job than we ever dare to expect in animal behaviour experiments (physicists expect higher precision because there is usually less statistical error in their measurements).

I also used all the same data to test the predictions of an alternative model, one which simply assumed that each colour has a 'value' for the animal, and that the animal allocates its choices in proportion to the colour's value. The two models gave similar predictions, so that if one is right the other one can't help being nearly right. But the Drive Threshold Model was consistently more accurate in predicting the observed result. The 'colour value' model consistently overestimated $P_{BestWorst}$. The 'colour value' model was falsified. The Drive Threshold Model triumphantly survived the attempt to falsify it, and indeed its predictions were (with the exception of the one experiment) remarkably accurate.

Does this good performance of the model really mean that there is something equivalent to a fluctuating 'drive' in the chick's head, crossing 'thresholds', and that something equivalent to tossing a coin happens when the drive is above more than one threshold? Well, Popper would say that the model survived a strong attempt to disprove it; but that says nothing about what the 'drive' and the 'thresholds' actually correspond to in the language of nerves and synapses. It is at least an interesting thought that you can make inferences about what is going on inside the head without cutting it open.

The same method of imagining a model and testing its predictions has proved enormously productive in many branches of science. In genetics, for example, you can infer the existence of chromosomes as one-dimensional linear sequences of genetic code without ever looking down a microscope, using only the data from breeding experiments. You can even work out the order in which the genes are arrayed along the chromosomes, and how far apart genes are from each other, entirely by imagining what might be the case and testing predictions in breeding experiments. As with my experiments on solidity and shading, I think of my Drive Threshold Model as an illustrative example of the *kind* of thing that can be done with a model, rather than as a conclusive discovery of what is really going on inside a chick's head.

I elaborated the Drive Threshold Model in various directions (that's also something that is supposed to happen according to Popperian philosophy) and tested nine predictions in all, with good success. One of these elaborations of the model, as I mentioned above, was an attempt to explain the exact timings of pecks ('samplings' of the position of the 'drive' relative to the 'thresholds'). The predictions of this model stood up well against data on black-headed gull chicks from my colleague and close friend Dr Monica Impekoven, a visitor to Bevington Road from Switzerland. We published a joint paper on this work.[1]

Another elaboration of the model, which I published as the 'Attention Threshold Model',[2] was an attempt to probe more deeply the 'penny tossing' of the original Drive Threshold Model: the indiscriminate choice of target when more than one threshold is exceeded. Briefly, I suggested that chicks attend to one dimension at a time – colour, shape, size, texture etc. – and in a definite order. Each of these attention systems has its own version of the Drive Threshold Model. The chick attends to the first dimension – say, colour. If the colour system's drive/threshold delivers a definite choice, the chick goes for the preferred colour, say Blue. But if the colour system's verdict is a 'penny toss' the chick switches its attention to something else, say shape, and ignores colour. From the colour system's point of view, choosing by shape is equivalent to choosing at random. But of course it is not random from the shape system's point of view. This trickle-down process continues through all the attention systems. If all else fails, the equivalent of a 'penny toss' is something like 'choose the nearest one'. The Attention Threshold Model yielded a series of further predictions (making nine predictions in all), which I tested with success.

Again, as with the solid shading experiments: is it possible that a

[1] R. Dawkins and M. Impekoven, 'The peck/no-peck decision-maker in the black-headed gull chick', *Animal Behaviour*, 17 (1969), pp. 243–51.
[2] R. Dawkins, 'The attention threshold model', *Animal Behaviour*, 17 (1969), pp. 134–41.

version of the Drive Threshold Model might apply to humans? I searched the scientific literature and found that several psychologists had done pairwise preference tests on humans. Their motivation was different from mine, but I could use their published results. There are various reasons why a psychologist might present a range of choices in all possible pair combinations: to test an idea in Voting Theory, for instance. Instead of offering a three-way choice between Conservative, Liberal and Socialist, with either winner-take-all or rank-order voting, a pollster might investigate the benefits of pairwise testing: 'How would you vote between Conservative versus Liberal (if there were no other choice), between Liberal versus Socialist (given no other choice), and finally between Conservative versus Socialist?' Anyway, for whatever reason, psychologists have presented humans with choices in all possible pairwise combinations. I was therefore able to feed their measurements of Best versus Medium, and Medium versus Worst, into my formula, and test my model's prediction of Best versus Worst. The data came from a diverse set of studies: American students choosing handwriting samples, American students choosing vegetables, American students choosing bitter/sweet tastes, and Chinese students choosing colours. In addition, I was especially delighted to be able to use a big study of preferences for composers shown by the members of the Boston Symphony Orchestra, the Philadelphia Orchestra, the Minneapolis Symphony Orchestra and the New York Philharmonic. Overleaf is a graph pooling all the results from humans. Once again, if the Drive Threshold Model's predictions were perfect, the points would fall along the diagonal line. I must say I was pretty excited when I saw how closely the prediction was fulfilled. Predictions in behavioural biology just aren't usually fulfilled that precisely!

The orchestra study was large, and processing the data laborious. I discussed the problem with my Uncle Colyear, who was by then in the Oxford Forestry Department, lecturing and advising on statistical methods. He suggested that I should learn to program the university computer. He and his wife Barbara got me started and helped me to write a program for the composer preferences. And so began my forty-year time-wasting and soul-consuming love affair with computer programming, a love affair that is now happily over: I am still an intensive computer user, but I now leave the programming to professionals.

Back then in the mid-1960s there was only one computer in Oxford University: a brand new English Electric KDF9, less powerful than an iPad today, but state-of-the-art in its time and filling a large room. The programming language favoured by my uncle and aunt was K-Autocode, a British alternative to Fortran with a similar structure and grammar, and a similar tendency to foster bad programming practice (absolute jumps, for instance). At that time American computers used great stacks of punched cards (vulnerable to being dropped and irretrievably shuffled) and British ones used punched paper tape (spewed out into great spaghetti mounds on the floor, which then had to be rolled up and were vulnerable to tearing). Thank goodness those days are over. Thank goodness, too, that computers now communicate with us via a screen or loudspeaker, not on reams and reams of paper – and immediately, instead of after a 24-hour delay.

But in those days we knew no better, and I was captivated. I was entranced by the idea of pre programming a sequence of operations and chugging through them step by step with a pencil to check them, then unleashing them in the computer to be rerun at very high speed and thousands upon thousands of times. I had a terrible night when I dreamed that I was a computer running my program and I spent – as it seemed – the whole night going through repeat loop after repeat loop in my fevered brain. To be fair, conditions that night were not ideal for sleeping anyway. Along with several of the Bevington Road crew, I had been persuaded by my friend Robert Mash to spend the weekend hunting the Surrey Puma.

Ever since 1959, sightings of a mysterious large carnivore had been reported in the woods of Surrey in southern England. Dubbed the Surrey Puma, it had achieved the status of a minor Yeti-style myth, and a group of us joined together to spend a weekend in May 1966 trying to find it. The newspapers heard about the scheme and, with news material getting thinner as the summer Silly Season approached, the *Observer* printed a photograph of me in a British Empire pith helmet (solar topee) of the kind that I had worn as a child. I have forgotten where my companions pitched their tents, but my assigned role was to spend the night in a sleeping bag, out in the open under the stars, surrounded by great haunches of raw meat. I was allocated a camera with a flash, and my marching orders were to photograph the puma if it should come to take the meat – or me, I suppose. I didn't sleep peacefully, to say the least, so perhaps it was not surprising that my computer nightmare should have come on that particular night. Dawn came as a great relief to me and my companions, and a dreamily misty dawn it was (as you can see from the photograph reproduced in the picture section). We never did find the Surrey Puma, and it may be significant that sightings went on being reported as late as 2005, which would seem to suggest that the Surrey Puma lived more than twice as long as the maximum recorded lifespan of its species, even in captivity.

My computer habit moved on from the KDF9 to a smaller but more accessible computer. Oxford Zoology had acquired a dynamic new professor (head of department in the Oxford-speak of the time) in succession to the genial and slightly batty-looking Sir Alister Hardy. The stiff new broom from Cambridge was 'Laughing John' Pringle (one of those ironic nicknames, like 'Lofty' for a very short man), and the department was thrown into a whirlwind of modernization. In one way after another, dear old Alister Hardy's dear old department was 'upgepringled', no doubt for the better. Among the more exciting instances of upgepringleheit was the arrival from London of an equally dynamic group of X-ray crystallographers (think Watson/Crick, but with protein molecules instead of DNA). And, most exciting for me, they brought with them their own computer, which I was allowed, by its friendly custodian Dr Tony North, to use at night when it wasn't needed for number-crunching the patterns of X-ray scattering by crystals. An Elliott 803, it was even more primitive by modern standards than the KDF9, but it had the enormous advantage that I was allowed to get my hands on it.

This was the time when I became fully aware of the addictive lure of computers. I really did literally – and frequently – spend all night in the warm, glowing computer room, entangled in a spaghetti of punched paper tape, which must have resembled my insomnia-tousled hair. The Elliott had the charming habit of beeping an acoustic rendering of its inner processing. You could listen to the progress of your computation through a small loudspeaker which hummed and hooted a rhythmic serenade, doubtless meaningful to Dr North's expert ear but merely companionable to my nocturnal solitude. Nocturnal dalliance with computers in youth is characteristic of those – now called geeks – whose love affair with computers lasted longer (and more profitably) than mine: Bill Gates, to name one. With hindsight I cannot say that my affair with the Elliott was productive. No doubt I obtained some valuable practice in the art of programming. But Elliott Autocode was not a language that could be used on any other computers, and my nocturnal geekery, though diligent and very hard work, bore the same relation to serious programming as my tootling in the Oundle music school bore to real music.

I gave a talk on my Drive Threshold Model at the 1965 International Ethological Conference in Zurich. For the talk, I built a physical model of my theory, incorporating a rubber tube filled with mercury which I jiggled up and down to represent fluctuating 'drive'. The rubber tube was attached to the bottom of a vertical glass tube, into which were let three electrical contacts at different depths, representing 'thresholds'. Mercury is an electrical conductor, so when the jiggling column hit any of these contacts (the 'drive' exceeded the 'threshold') a circuit was completed. Obviously, if mercury was in contact with any electrode, it was automatically in contact with all lower electrodes too, which captured the major assumption of the model. I implemented the rules of the model by means of a noisy system of clattering electromechanical relays switching on coloured lights to represent pecks at different colours. The whole Heath Robinson[1] affair was calculated to bring the house down, just as, at an earlier Ethological Conference in Oxford, a spoof hydraulic simulation devised by Desmond Morris, Aubrey Manning and friends reputedly had. How I managed to transport it from Oxford to Zurich evades my

---

[1] American: Rube Goldberg.

memory, and indeed my comprehension. There's not a chance that today's airport security would allow anything remotely like it through, bristling as it was with amateurishly soldered wires, relays, batteries and mercury.

Alas, just as I was about to go on the big stage for my first ever conference performance, something went wrong and my contraption wouldn't work. In a sweat of panic and unable to think straight, I was kneeling on the floor outside the theatre, frantically tinkering, when I suddenly became aware of an amused Austrian accent barking out peremptory orders at great speed behind me. The rapid-fire, stuttering voice told me exactly what to do. As in a dream I obeyed – and it worked. I turned to look at my saviour, and beheld Wolfgang Schleidt, whom I hadn't previously met although I knew who he was. Without any prior knowledge of what my infernal machine was supposed to do, this rising star of continental ethology had come upon my panic, instantly sized up the problem and dictated the solution to me. I have been grateful ever since to Dr Schleidt, who had, as I later learned without surprise, a reputation for technical ingenuity. I bore my strange device up into the theatre and at the end of my talk its spluttering coloured lights and air of Heath Robinson amateurishness received a reception little short of an ovation. Thank you, Wolfgang Schleidt: and for more than saving my blushes. For in the audience that day was the handsome figure of George Barlow, rising star of American ethology, and he was sufficiently impressed by my talk to get me invited to become an assistant professor at the University of California at Berkeley, without an interview or submission of a curriculum vitae: my first proper job.

But that was to come later. Meanwhile, back in Oxford, Niko Tinbergen had a sabbatical leave in 1966 and he invited me to take over the undergraduate lectures on animal behaviour for that year. He offered me his lecture notes, but I decided instead to develop my own lectures from scratch. Because this was the first course of lectures I ever gave, I wrote them out pretty fully. I thought I had lost them long ago but, to my surprise, they turned up in a cardboard box in the basement of my house while I was writing this memoir, and it was quite interesting for me to read them forty-six years on – especially the lecture on social behaviour, because it so clearly demonstrates both the central message and the style of *The Selfish Gene*, despite having been written a full ten years before that book.

In 1964, the *Journal of Theoretical Biology* had published two long and rather difficult mathematical papers by W. D. Hamilton, a young graduate student at the University of London whom none of us knew at the time, although he was later to become a close colleague. Mike Cullen characteristically recognized the importance of Hamilton's papers before almost anyone else in the world except John Maynard Smith, and he expounded them one evening to the Bevington Road group. Mike's enthusiasm was infectious, and I was immediately fired up, to the extent that I wanted to explain Hamilton's ideas to the undergraduates in my stand-in lectures on animal behaviour.

Hamilton's theory, now often labelled 'kin selection' (Maynard Smith's name, not Hamilton's own), follows directly from the neo-Darwinian 'Modern Synthesis' – directly, in the sense that kin selection is not an extra, not an addition bolted on to the neo-Darwinian synthesis: it is a necessary part of the synthesis. You cannot divorce kin selection from neo-Darwinism, any more than you can divorce the Pythagorean theorem from Euclidean geometry. A field biologist attempting to 'test' kin selection is in the same position as Pythagoras setting forth with a ruler seeking triangles to measure.

The neo-Darwinian synthesis, as opposed to Darwin's own version of his theory, is centred on the gene as the unit of natural selection. Genes are discrete entities that can be *counted* in a population, more or less ignoring the fact that they are in fact sitting in the cells of organisms. Each gene has a *frequency* in the 'gene pool', which is approximated by the number of reproductive individuals possessing it. Successful genes are those whose frequency increases at the expense of unsuccessful alternatives, which decrease in frequency. Genes that cause an animal to be good at caring for its offspring tend to increase because they are carried in the bodies of the offspring cared for. Hamilton realized (as Fisher and Haldane had too, sort of, but they didn't make much of it) that offspring are not the only category of relatives who share genes, and who therefore might be beneficiaries of evolved caring.

Hamilton derived a simple rule (now called Hamilton's Rule): any gene 'for' altruism towards kin will tend to spread through the population, if the cost to the altruist C is less than the benefit B to the recipient devalued by the degree of relatedness r between the two. The degree of relatedness r is a proportion (i.e. a number between 0 and 1) which Hamilton showed how to calculate (its exact meaning is hard, though not impossible, to explain intuitively).[1] Between full siblings, r is 0.5. Between uncles and nephews it is 0.25, between first cousins 0.125. Hamilton had a special interest in the social insects, and he made brilliant use of his theory of kin selection in explaining how ants, bees, wasps and (in a rather different way) termites evolved their remarkable habits of social altruism.

A typical underground nest of ants is a factory for propagating genes and spreading them around the countryside. The genes are churned out from the factory packaged up inside the winged bodies of young queens and males. These flying ants (which you might not realize are ants because of their unfamiliar wings) erupt from holes in the ground and fly up to mate on the wing. During her mating flight, each female (young queen) collects a lifetime's supply of sperm, which she will store inside her body and eke out over the course of a long life. Laden with sperm, the mated female flies off and settles down to dig a

---

[1] The clearest explanation is given by my Oxford colleague and sometime graduate student Professor Alan Grafen, 'A geometric view of relatedness', in R. Dawkins and M. Ridley, eds, *Oxford Surveys in Evolutionary Biology*, vol. 2 (Oxford, Oxford University Press, 1985), pp. 28–89.

hole and found a new nest. In some species, she bites or breaks off her wings, as she will need them no more in her role as subterranean queen.

Most of her offspring will be sterile workers, but the important children from the point of view of gene propagation are the young (winged) queens and males. Workers (all female in the case of ants, bees and wasps; male and female in the case of termites) normally have no prospect of passing on their genes through offspring, and devote their efforts to feeding and caring for their fertile collateral relatives, young queens and males, their siblings or nieces, for instance. A gene that makes a sterile worker care for a sister who is destined to become a queen can pass into future gene pools, carried there in the body of the young queen. The caring behaviour may never be expressed by the young queen herself, but the gene for that behaviour will be passed on to her worker daughters, who will consequently care for young queens and males who can pass it on.

The social insects are just a special case. Hamilton's Rule applies to all animals and plants, whether or not they in practice care for relatives. If they don't care for relatives, the reason will be that the economic costs and benefits in Hamilton's Rule (the *Bs* and *Cs*) don't pan out in such a way as to favour it, in spite of the fact that the coefficient of relatedness *r* may be high. And – a point that is often misunderstood, even by professional biologists – individuals care for their offspring for the same reason that elder siblings care for younger siblings (when they do): in both cases they share the genes for caring.

As I said, I was fired with enthusiasm when Mike Cullen introduced us to Hamilton's brilliant ideas, and I very much wanted to try my hand at explaining them in my own way, in the lectures that I was to give as understudy for Niko Tinbergen. I was diffident about departing so far from Niko's message and substituting my own rhetoric – about 'selfish genes' inhabiting a succession of mortal bodies, to be thrown away in the genes' relentless march into the future. Seeking

Natural selection acts directly on phenotypes, but it will affect
evolution only insofar as phenotypic ~~xxxxxxxxxxxxx~~ differences
are correlated with genetic differences. The important effect of
natural selection is therefore on <u>genes</u>.

Genes are in a sense immortal. They pass through the generations,
~~xxxxxxxx~~ reshuffling themselves each time they pass from parent to
offspring. The body of an animal is but a temporary resting place
for the genes; the further survival of the genes depends on the
survival of that body at least until it reproduces, and the genes
pass into another body. The structure and behaviour of the body
are to a large extent determined by the genes - the genes build
themselves a temporary house, mortal, but efficient for as long as
it needs to be. Natural selection will favour those genes which
build themselves a body which is most likely to succeed in ~~xxxxxx~~

lovely stuff

handing ~~xxxx~~down safely to the next generation, ~~xxx~~ a large number
or replicas of those genes.

To use the terms "selfish" and "altruistic" then, our basic
expectation on the basis of the orthodox neo-Darwinian theory of
evolution, is that <u>Genes will be "sefish"</u>.

~~Must this mean that individuals will be selfish? Not necessarily,~~
~~though it does mean that we must be very suspicious of expressions~~
~~like "the good of the species" There are two main ways in which~~
~~individual altruism~~

(This gives us the most important difference between individuals and
social groups. If an individual body is a colony of cells, it is
a very special kind of colony, because all those cells are genetically
identical. Every ~~xxxxxxxxxxxxx~~ somatic cell, muscle, bone, skin,
brain etc., contains the same complement of genes. Furthermore the
reproduction of all the genes in these somatic cells is limited to
the life-span of the body. Only the genes in the germ cells ~~will~~ *may*
survive. The other cells are built by the genes simply to ensure the
survival of the ~~xxxxxxxxxx~~ identical genes in the germ cells.
In say a ~~flock~~ colony of gulls, the individual birds all contain
different ~~xxxx~~ sets of genes (except identical twins), and because
of the arguments given above, we shall have to think very carefully
about whether we should expect altruism between individuals. Only
in the social insects where the workers are sterile and very closely
related, do we have a social group that is really comparable with
the many-celled body. We will return to this later.)

If genes are selfish then, how can individuals evolve altruism?

reassurance, I showed my typed-out lecture to Mike Cullen, and seeing again his handwritten *marginalis* today reminds me that I was hugely encouraged by it at the time (see the facsimile opposite). It was Mike's 'lovely stuff' that emboldened me to persist with my plan to lecture on this topic and in this style. And I suppose this could be said to be the moment of conception of *The Selfish Gene*, born ten years later. My lecture notes even contained the phrase 'genes will be selfish'. I'll return to this later when I come to the book itself.

In the summer of 1967, in the tiny Protestant church of Annestown on the south coast of Ireland, where her parents had a holiday cottage, I married Marian Stamp, a member of Niko Tinbergen's group of graduate students, who later became his successor but one as Oxford's Professor of Animal Behaviour and the world's leading authority on the experimental science of animal welfare. I was by now committed to taking up the offer of an assistant professorship at the University of California at Berkeley. Niko was confident of Marian's ability to continue her doctoral research there, with a minimum of long-distance supervision from him, and indeed his confidence was well justified. We had a brief honeymoon driving around Ireland in a hired car. Marian had to drive because I had forgotten my licence, and we had an awkward moment when the car hire clerk discovered she was a 'graduate' (it seems that graduates had a poor track record). Almost immediately after the honeymoon we set off for San Francisco, where we were met at the airport by the ever-kind George Barlow. A new life in the New World had begun.

WEST COAST DREAMTIME

BERKELEY in the late 1960s was politically seething, and the politics of Telegraph Avenue, and Haight-Ashbury across the bay in San Francisco, were to dominate our two years there. Lyndon Johnson, who might otherwise have been remembered as a great reforming president, was mired in the disaster that was the Vietnam War – inherited from Kennedy. Just about everyone in Berkeley was against the war, and we joined them – in marches in San Francisco, in tear-gassed parades in Berkeley, in demonstrations, classroom disruptions and sit-ins.

I am proud of my part in protesting against American involvement in Vietnam, proud of having worked hard in the anti-war campaign of Senator Eugene McCarthy, less proud of some of the other political movements in which I was involved. The most memorable of these concerned the surreal episode of the 'People's Park' (fictionalized by David Lodge as the 'People's Garden' in his campus novel *Changing Places*). The People's Park campaign was an attempt (ultimately successful, as I discovered when I revisited Berkeley on a filming trip recently) to take over for public recreation a piece of waste ground owned by the university and intended for building. With hindsight it was a trumped-up excuse for radical political activism for its own sake, trumped up by anarchist student leaders cynically manipulating the gentle 'flower-power' 'street people'. The radical student leaders and the infamous Governor Ronald Reagan ('Ronald Duck' in David Lodge's novel) gleefully played into each other's hands, each mining the situation to enhance their following among their respective constituents, and each probably knowing exactly what they were doing. And I, together with most of the younger faculty of the university, played right into their hands. We demonstrated, sat in, ran from the tear gas, wrote outraged letters to the newspapers (my first letter to *The Times* was on the subject) and cheered as the street people stuck flowers down the rifle barrels of the bewildered and rather scared young National Guardsmen. Honesty compels me to admit a frisson of exhilaration – of which I am now quietly ashamed – at having been tear-gassed and (very slightly) endangered.

I try to peer into my own state of mind in my twenties in Berkeley as honestly as I can. I think what I see there is a kind of youthful excitement at the very idea of rebellion: a Wordsworthian 'Bliss was it in that dawn to be alive / But to be young was very heaven'. A student called James Rector was shot dead by an Oakland policeman. It was right to march in protest against that, and with hindsight that seems to have justified, in our minds, our decision to march for the People's Park in the first place. But of course it didn't justify it at all, not in itself. The decision to march for the People's Park required completely separate justification.

We, the younger faculty, convened meetings where we tried to bully our colleagues into cancelling their lectures in solidarity with the activists – and I use the word 'bully' advisedly, for I have seen the same thing more recently on the internet in the form of 'cyberbullying' by radical activists powerful enough to act as a kind of thought police, just as I saw the same thing at school when willing accomplices would rally around a playground bully. I remember with particular regret a faculty meeting at Berkeley when a decent older professor was reluctant to cancel his lecture and we tried to vote to force him to do so. With remorse I salute his courage, and that of an even older professor whose hand was the sole one raised in support of his colleague's right to fulfil what he perceived as his duty to give his scheduled lecture. As with Aunty Peggy, as with the Chafyn Grove equivalent, I should have stood up against the bullies. But I didn't. I was still young, but not all that young. Should have known better.

Mention of radical politics and the street people brings forth a revealing memory, revealing of a sea-change in social mores. I was walking along Telegraph Avenue, axis of Berkeley's beads-incense-and-marijuana culture. A young man was walking ahead of me, dressed in the insignia of the flower-power generation. Every time a young woman passed him, walking in the opposite direction, he would reach out and tweak one of her breasts. Far from slapping him, or crying 'Harassment!', she would simply walk on by as if nothing had happened. And he would proceed to the next one. Today I find this almost impossible to believe, but it is a very secure memory. His demeanour did not appear especially lascivious, and his action was evidently not taken by the young women as the gesture of a male chauvinist pig. It seemed all of a piece with hippiedom, with the laid-back, peace-and-love atmosphere of sixties San Francisco. I am very glad to say that things have changed. Today's counterparts in age and class of that young man, and the young women he molested (as we should now say), would be among those most strongly outraged at behaviour which was then the norm for that age, class and political persuasion.

In spite of all the politics, I did an adequate job as a junior (indeed, exceptionally young) assistant professor. George Barlow and I shared the lectures on animal behaviour, and I included the 'selfish gene' lecture that I had initiated at Oxford. I like to think that the students at Oxford and Berkeley in the late sixties may have been the very first undergraduates in the world to hear of the new ideas that were to become fashionable, in the seventies and thereafter, as 'sociobiology' and 'selfish genery'.

Marian and I were made to feel very welcome at Berkeley, and we made good friends there. As well as George Barlow, these friends included David Bentley the neurophysiologist, Michael Land, now the world's leading authority on eyes throughout the animal kingdom, and Michael and Barbara MacRoberts, who later came to Oxford as spirited additions to the Bevington Road circle, as did the gently sardonic David Noakes, who was George Barlow's leading graduate student during my Berkeley years. George hosted a weekly ethology seminar for interested graduate students at his house in the Berkeley Hills, and those evening meetings recaptured for Marian and me something of the wonderful atmosphere of Niko's Friday evenings at Oxford.

I had never been to America before, and I did find some things bewildering. At my first meeting of the Zoology faculty, everyone spoke almost entirely in numbers. Who's doing 314? No, I'm doing 246. Nowadays the English-speaking world knows that Xology 101 means (sometimes patronizingly or even derisively) a freshman's introduction to Xology. But all that numerology was perplexing to me when I first arrived. And who, today, doesn't understand the verb 'to major'? But I recall reading an American campus novel and getting a little fed up with the twittering of sophomores and juniors and seniors when, like a breath of fresh air, 'An English major came into the room.' Aha, I thought, my mind immediately filled with visions of riding breeches and moustaches, a real character at last.

Marian and I both worked hard at our research. And we talked and talked and talked to each other about our shared scientific interests, on walks in Tilden Park up in the Berkeley Hills, on drives around the beautiful California countryside, at meals, on shopping expeditions over the Bay Bridge to the city of San Francisco, all the time. The atmosphere of our discussions was that of a mutual tutorial, each learning from the other, exploring arguments step by step, moving one step backwards, two steps forwards. The mutual tutorial is something I now strive to achieve in public discussions with colleagues, often filmed for my website or put out on DVDs. Those discussions with Marian were to be the basis of the joint experiments that we did later, after returning to Oxford.

My research at Berkeley was a continuation of my work on chick pecking. My doctoral research had been very Popperian. It made precise predictions about total quantities of choices made in a fixed time. But the model had always begged for more exact observational testing, using precise sequences of pecks as they happened, rather than total numbers of pecks per minute. In Berkeley I turned to the exact sequence, building a new apparatus which, unlike my Oxford one, was capable of recording exactly when each peck happened, rather than just counting pecks per minute. I also increased the peck rate by rewarding each peck with a blast of infra-red heat, which the chicks liked. They were rewarded equally, regardless of which key they pecked, but they still showed colour preferences and they still seemed to be choosing on the basis of the Drive Threshold Model. The pecks were recorded on magnetic tape using a sophisticated and expensive piece of equipment that had been built for George Barlow, known as the Data Acquision System – so called because of a typographical error in the word 'Acquisition' on the label.

One simple expectation of the Drive Threshold Model is that there should be long runs of pecks at the preferred colour (when the drive was above the threshold for that colour only), interspersed with runs of indifference (when the drive was above two thresholds). There should never be significant runs of pecks at the less preferred colour. Following the Attention Threshold Model, I expected that indifference to *colour* really meant preference for one *side*. Since the apparatus was programmed to present each colour to alternate sides, changing after every peck (with occasional random variations), I predicted sequences such as you see in the picture on this page, which represents real data from one particular experiment, and seems to confirm the prediction very nicely.

Of course, this picture is no more than an anecdote, one out of many experiments. I did statistical analyses to substantiate this prediction and several other predictions, using data from large numbers of experiments. The predictions of the Attention Threshold version of the Drive Threshold Model were upheld.

Some time during our second year at Berkeley, Marian and I were visited by Niko and Lies Tinbergen. Niko wanted to persuade us to return to Oxford, where he had obtained an attractive research grant to offer me, and where Marian could write up her doctoral research, which, as Niko could see, was going well at Berkeley. The Tinbergens returned to Oxford, leaving us to think about the offer. We decided to accept it, but meanwhile Niko had written of a new opportunity. Oxford had decided to appoint a new university

Successive columns represent sequences of pecks. Open circles represent preferred colour. Note how runs of pecks at the preferred colour (open circles) alternate with runs of pecks at one side or the other – perhaps the nearest side – amounting to runs of alternation between colours.

From M. and R. Dawkins, 'Some descriptive and explanatory stochastic models of decision-making', in D. J. McFarland, ed., *Motivational Control Systems Analysis* (London, Academic Press, 1974), pp. 119–68.

lecturer[1] in animal behaviour linked to a fellowship at New College, and Niko wanted me to apply. This teaching job would not preclude the research grant he had earlier promised me. I agreed to apply for the lectureship, and Oxford flew me over for the interview.

It was a magical trip, with what seemed like all before me. Music stamped the memory: Mendelssohn's Violin Concerto, which I listened to on the plane, spellbound by the Rocky Mountains below and by exciting prospects ahead. Oxford put on its very best performance, which is the Maytime blossoming of cherry and laburnum all along the Banbury Road and Woodstock Road. New College, too, played its golden fourteenth-century part and I was happy, my exuberance not dimmed when I was greeted on arrival by the news that Colin Beer, former member of the Oxford ABRG and now a professor at Rutgers University in New Jersey, had put in an unexpected late application for the lectureship. Even the fact that Niko had excitedly switched his allegiance from me to Colin didn't upset my optimistic mood. If Niko had decided that Colin was a better bet, that was good enough for me. I would still have the research position and, as I told the interviewing committee, if Colin were there in Oxford too, so much the better. They did indeed give the job to Colin, and I took up the research grant.

[1] The American equivalent would be 'assistant professor going on associate professor'.

COMPUTER FIX

MARIAN and I left Berkeley in 1969 with mixed emotions, and it has remained for me a place of magic and pilgrimage: a dreamtime of lost youth, of clever and friendly colleagues, of clear, bright sunshine alternating with cooling mist over the Golden Gate, of the dawn-fresh scent of pine and eucalyptus, of flower children with decent and sincere, if naive, liberal values.

We crated and shipped our few belongings from our Berkeley apartment and drove right across the continent in our old, cream-coloured Ford Falcon station wagon, thickly encrusted with anti-war slogans and Eugene McCarthy election stickers, to New York. By prior arrangement, we sold the Ford on the quayside (amazingly, our pre-arranged customer, who had made his own laid-back Berkeley-style way to New York, showed up on time), boarded the liner *France* for Southampton, and prepared to resume our life at Oxford with many of our old friends still there and Colin Beer newly arrived. In the event, Colin preferred to spend his time in New College and was scarcely seen in the department, much to everyone's disappointment. He stayed only a year. Danny Lehrman – the same Daniel S. Lehrman whose theoretical critique had so influenced my doctoral thesis – had shrewdly kept Colin's position at Rutgers warm for him, and when it became clear that Oxford couldn't find a position in medieval French to match the professorship his wife held in America, Colin decided to return. Once again, the lectureship in animal behaviour was advertised, once again the long-suffering New College agreed to associate a fellowship with it, and once again Niko urged me to apply. Along with a shortlist of others I was again interviewed by two committees: a university committee chaired by Laughing John Pringle, and a college committee chaired by the genuinely laughing, almost preternaturally genial Warden, Sir William Hayter, former British Ambassador in Moscow.

This time I really wanted the job, and this time I got it. The news came as Marian and I were anxiously waiting with friends in an Indian restaurant in Oxford. Suddenly, we heard the sound of Mike Cullen's motor scooter as it pulled up outside. Mike burst into the restaurant, wordlessly pointed both index fingers at me, and vanished as quickly as he had arrived. I had got the job. With hindsight I don't think I should have got it at the time, given that the chief competition was the wildly brilliant Juan Delius, although I like to think that I grew into it and was worthy of it in the end. Juan was a dear friend and mentor, an immensely clever, knowledgeable and funny German-Argentinian. He once defined Argentinian humour for me: 'They enjoy slapstick, but if somebody slips on a banana skin it's only *really* funny if he breaks his leg.' The board at 13 Bevington Road was frequently adorned with wonderful notices in Juan's unique brand of English: 'What bastard has absconded my oles?' (Who has taken my stencil for drawing circles of different sizes?)

The life of a tutorial fellow of an Oxford college is in many ways a charmed one. I got a room in a glowing, oolitic limestone medieval building surrounded by famously beautiful gardens; a book allowance, a housing allowance, a research allowance; and free meals (though not free wine, contrary to envious rumours) in the stimulating and entertaining company of leading scholars of every subject except my own. The stimulating scholars of my own subject were to be found in the Zoology Department – where I spent the majority of my time.

I was introduced to the strange world of high-table conversation. After dinner there would sometimes be occasion to bring out the Senior Common Room Betting Book – either to record a new wager or to browse the old ones, all written in the same affected style as high-table conversation itself. Here's a brief sample, going back to the 1920s when the most assiduous betting man was the eccentrically brilliant G. H. Hardy, whose Lewis-Carroll-like mathematician's sense of humour seems to have infected his colleagues:

> (7th Feb 1923) The Subwarden bets Prof Hardy his fortune till death to one halfpenny that the sun will rise tomorrow.

> (6.8.27) Prof Hardy bets Mr Woodward 10,000 to 1 in halfpennies that he (Prof Hardy) will not be the next President of Magdalen, and Mr Woodward bets Prof Hardy 1 to 5,000 that he (Mr Woodward) will not be the next President of Magdalen.

> (Feb 1927) Professor Hardy bets Mr Creed 2/6 to 1/6 that the New Prayer Book will go phut. Mr Smith, Mr Casson and Mr Woodward to adjudicate if necessary.

I'm amused that so obvious a value judgement could be subject to a bet. No wonder an odd number of adjudicators was needed.

Another bet even leaves the size of the stake up to later judgement:

> (Dec 2 1923) Professor Turner bets the Steward of SCR a large sum that it would be a good thing to have a copy of the ABC (London) Railway Guide in SCR (Won by Prof Turner, A.H.S.)

> (15 Feb 1927) Mr Cox bets Professor Hardy 10/- to 1/- that Rev. Canon Cox ('Fred') will not be the next Bishop of Nyasaland.

I love that parenthetic 'Fred'. Unfortunately the result of this bet was not recorded. I'd like to know whether 'Bishop Fred' presided over the see of my early home country. In failing to settle the question for me, Google did turn up the fact that a nineteenth-century Bishop of Nyasaland was Charles Alan Smythies – very probably related to the seven generations of my Smythies ancestors who were vicars.

(11th March 1927) Mr Yorke bets Mr Cox 2/6d that no verse occurs in the Gospel according to St Matthew the literal interpretation of which justifies or advocates self-castration. Won by Mr Cox.

(October 26th 1970) Professor Sir A Ayer bets Mr Christiansen that the Chaplain will be unable, if challenged without warning, to repeat twelve of the thirty nine articles to be found in the Book of Common Prayer. The stake to be one bottle of claret.

(24th Nov 1985) The Chaplain bets Dr Ridley a bottle of claret that Dr Bennett will be wearing a clerical collar at dinner on the occasion of the visit of the Bishop of London. (Chaplain won.)

(4th August 1993) Mr Dawkins bets Mr Raine £1 that Bertrand Russell married Lady Ottoline Morrell. Adjudicator Mlle Bruneau. (Dawkins lost and paid, 20 years late.)

Bets like the last one can't happen any more because it is so trivially easy for everyone to check such factual questions on their smart-phones without rising from their Senior Common Room armchairs. Even then, it was scarcely necessary to appoint an adjudicator for a purely factual matter.

Back to 1970, when I was twenty-nine and newly returned to Oxford. The singing Elliott had gone the way of all silicon, but Moore's Law and the research grant that had lured me back to Oxford the previous year made it possible for me to have my 'own' computer, a PDP-8, which exceeded the Elliott in every respect except physical size and price. Also in accordance with Moore's Law (which was already going strong in those days), it was functionally much smaller yet physically larger than a modern laptop, and ludicrously it had a log book in which you were supposed to record every time you switched it on (of course I didn't). It was my pride and joy and a valued resource – together with me as sole programmer for everybody in 13 Bevington Road (which took its toll on my time). Now my addiction to computers could really take off, and I no longer had to indulge it nocturnally, as during my shameful affair with the Elliott 803.

Previously I had used only high-level compiler languages – human-friendly languages, which the computer translates into its own binary machine language. But now, in order to use the PDP-8 as a research tool, I had to master its 12-bit machine language, a task into which I threw myself with zest. My first machine-code project was the 'Dawkins Organ', a system for record-ing animal behaviour – equivalent to George Barlow's 'Data Acquision' apparatus but much much cheaper. The idea was to make a keyboard which an observer could use in the field, pressing buttons to indicate actions by an animal. Key-presses would be recorded on a tape recorder, which would later automatically tell the computer exactly when each action by the animal occurred.

My keyboard literally was a makeshift electronic organ, with each key playing a different note (inaudible except to the tape recorder). This part would be easy to make. The box would contain a simple two-transistor oscillator, the pitch of whose note was tuned by a resistance. Each key on the organ would connect a different resistor and hence play a different note. The observer was to take the organ into the field and watch an animal's behaviour like a work-study officer, pressing a specific key for each behaviour pattern. A tape recording of the sequence of notes would then constitute a timed record of the animal's behaviour. Theoretically, a person with a good ear listening to the tape could detect which key had been pressed, but this wouldn't be helpful. I needed to cast the computer in the role of person-with-good-ear. It could have been done electronically, with a series of tuned frequency-detectors, but that would have been an expensive hassle. Could the same feat – perfect pitch sensitivity in the computer – be achieved in software alone?

I was discussing the problem with my computer guru of the time, Roger Abbott, a clever engineer (and coincidentally organist) employed on the large research grant of Professor Pringle. Roger came up with an inspired suggestion. Every musical note has a characteristic wavelength which signifies its pitch. Computers are – and were, even in those days – so fast that the interval between wave crests within a musical note could be measured in hundreds of program cycles. Roger suggested that I should write a machine-code program to time the intervals between wave peaks: write, in other words, a little routine to act as a high-speed clock, counting how many jumping-back program loops it could walk through before being interrupted by the next wave crest (which, when averaged over lots of wave crests, tells it the pitch of the note). When a note ended (when more than a critical time elapsed since the last wave peak) the computer should make a note of the time and then wait for the next organ note. The computer's clocking loop, in other words, would be used not only to recognize the pitch of a musical note but also, on a hugely longer timescale, to measure the passage of time between notes.

Having got this central routine working, the rest was just a matter of slogging through the writing and debugging of a user-friendly program. This took rather a long time, but it ended successfully. The Dawkins Organ was a viable product. The user of the organ began each session by playing a scale on the tape – all the notes on the organ in ascending order of pitch. The taped scale would then be used to 'calibrate' the software – 'teach' the computer the repertoire of notes it would be asked to recognize. After the calibration scale was ended (by hitting the first note for the second time), all further notes on the tape would designate behavioural events. This calibration system had the advantage that the organ did not have to be carefully tuned. Any set of notes that were sufficiently distinct from each other would do, because the computer quickly learned which notes to listen out for.

So, when the tape was brought home and played into the computer, the computer knew exactly what the animal had done, and when. The nucleus of the program was the timing loop, but it was embedded in a substantial quantity of code to punch out, on paper tape, the names of all the behaviour patterns and the exact times when they occurred.

I published a paper on the Dawkins Organ,[1] and made the software available free of charge. Over the next few years Dawkins Organs were used by numerous members of the Oxford ABRG, and by some ethologists elsewhere in the world, for example in the University of British Columbia.

My addiction to machine code programming took me in a downward spiral. I even devised my own programming language, BEVPAL, with its own programming manual, a somewhat otiose exercise since the language was used by nobody except myself and, briefly, Mike Cullen. Douglas Adams amusingly satirized computer addiction of exactly the kind that hit me. The target of his satire was the programmer who had a particular problem X, which needed solving. He could have written a program in five minutes to solve X and then got on and used his solution. But instead of just doing that, he spent days and weeks writing a more general program that could be used by anybody at any time to solve all similar problems of the general *class* of X. The fascination lies in the generality and in the purveying of an aesthetically pleasing, user-friendly product for

---

[1] R. Dawkins, 'A cheap method of recording behavioural events for direct computer access', *Behaviour*, 40 (1971), pp. 162–73.

the benefit of a population of hypothetical and very probably non-existent users – not in actually finding the answer to the particular problem *X*. Another symptom of this kind of geekish addiction is that every time you solve a local problem and make the computer jump through yet another hoop, you want to rush out into the street and drag someone in to show them how elegant it is.

The productive camaraderie that a small building like 13 Bevington Road fosters came to an end around this time, and the animal behaviour group moved to the new zoology/psychology building, the huge, battleship-like horror on South Parks Road, then informally known as HMS Pringle after the ambitious Linacre Professor who persuaded the university authorities to build it – having failed to cajole them into building a pencil-thin skyscraper that would have disastrously overtopped Matthew Arnold's dreaming spires. I have mixed feelings about my part in later getting HMS Pringle officially named the Tinbergen Building, for it is widely deplored as the ugliest building in Oxford. It won an architectural award from the Concrete Society – enough said.

Around this time I published a short paper in *Nature*.[1] Every day hundreds of thousands of our brain cells die, and this was upsetting to me even at the age of twenty-nine. My Darwin-obsessed brain sought comfort in the idea that if the cell deaths were non-random, such apparently wholesale slaughter might be constructive, not purely destructive:

> A sculptor changes a homogeneous lump of rock into a complex statue by subtraction, not addition, of material. An electronic data processing machine is most likely to be made by connecting components up in complex ways, and then enriching the connexions to make it even more complex. On the other hand, it could be con-

---

[1] R. Dawkins, 'Selective neurone death as a possible memory mechanism', *Nature*, 229 (1971), pp. 118–19.

structed by starting with extremely rich, even random intercon-
nexions, and then carving out a more meaningful organization by
selectively cutting wires.

. . .

The theory proposed here may seem fanciful at first. Further
reflexion shows, however, that its lack of verisimilitude is mainly a
consequence of the highly improbable postulate on which it rests;
namely, that brain cells are decreasing in numbers at a prodigious
rate daily. Because this postulate, however far-fetched, is an established fact, the present theory is not suggesting anything very
implausible in addition; rather the reverse, as it makes the process
seem less wasteful. All that is at issue is whether neurones die at
random, or selectively in such a way as to store information.

A curious little one-off, this paper is perhaps mildly interesting as an
early example of the kind of theory that later became fashionable
under the name – coined a year later, so I obviously didn't use it –
'apoptosis'.

Marian soon got her doctorate, and we began to collaborate on
research projects growing out of the many discussions – mutual
tutorials – from our Berkeley days. We planned a study that would
exemplify, and clarify, one of the fundamental concepts of the ethological school of animal behaviour studies, the Fixed Action Pattern.

Lorenz and Tinbergen and their school thought that much of
animal behaviour consisted of a sequence of little clockwork routines
– Fixed Action Patterns (FAP). Each FAP was thought to be like a piece
of anatomy, just as much a part of the animal's bodily equipment as,
say, the collar bone or the left kidney. The difference is that collar bones
and kidneys are made of solid material, whereas the FAP has a time
dimension: you can't pick it up and put it in a drawer, you have to
watch it play out in time. A familiar example of an FAP would be the
pushing movement that a dog makes with its snout when burying a
bone. These movements are identically replayed even when the bone

is on a carpet and there is no soil in which to bury it. The dog really does look like a (charming) clockwork toy, although the exact direction of the movement is influenced by the position of the bone.

Every animal has a repertoire of FAPs, like one of those dolls that you wind up by pulling a string and which then utters a saying randomly plucked from a finite repertoire. Once initiated, whichever saying is chosen goes through to completion. The doll doesn't switch messages halfway through. The decision of which of a dozen sayings to produce is unpredictable but, once taken, the consequences of the decision are followed through predictably as clockwork. That was the FAP doctrine in which Marian and I, as Tinbergenian ethologists, had been brought up; but was it a true reflection of reality? This was the question we wanted to answer – or, to be precise, the question we sought to re-express in terms that might make it answerable.

In theory, one could write down the continuous stream of animal behaviour as a sequence of muscular contractions. But if the FAP theory was right, the predictability of behaviour would render it a laborious waste of effort to write down every muscular contraction, even were it possible to do so. Instead, all we should need to do is write down the FAPs, and the sequence of FAPs would – on an extreme interpretation – be a complete description of that particular animal's behaviour.

But this would only work if FAPs really were equivalent to organs or bones – if it were true, in other words, that each pattern occurs as a whole, not breaking off part-way through or mixing with another pattern. Marian and I wanted to find a way to assess the extent to which this proposition was true. Both our doctoral theses had been concerned – in our two ways – with decision-making, and it was natural for us to translate the FAP problem into the language of decisions. In that language, the animal takes a *decision* to initiate an FAP; but, once initiated, the FAP goes on to its conclusion, with no further decisions until the end. At that point the animal's behaviour stream would enter a period of uncertainty, pending the next decision to initiate (and complete) an FAP.

We chose to study drinking in chicks as our example, and we hoped it would be representative.[1] Drinking in birds (other than pigeons and doves, which suck) is an elegant glissando of a movement, and it certainly gives a subjective impression of being initiated by a discrete decision, after which it always goes through to completion. But could we back up our subjective impression with hard data?

We filmed a side view of our chicks drinking, and then analysed the behaviour frame by frame to see if we could measure its 'decision structure'. We measured the position of the bird's head in successive frames of film, then fed the coordinates into the computer. The idea was to measure the predictability of the next frame, knowing the position of the head in previous frames.

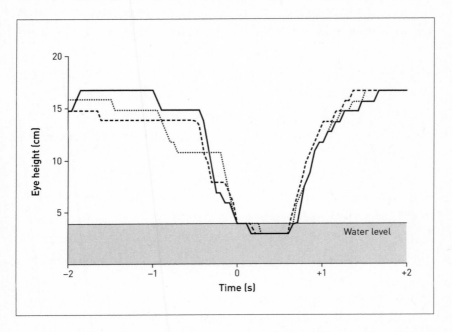

[1] R. and M. Dawkins, 'Decisions and the uncertainty of behaviour', *Behaviour*, 45 (1973), pp. 83–103.

The diagram shown on page 225 is a graph of eye height against time, for three drinks by the same chick, lined up (zero on the time axis) on the moment when the bill hit the water. You get a sense that from that moment on, indeed from just before it, the behaviour is stereotyped and predictable, but the early part of the downstroke is more variable and subject to decisions: decisions to pause and even (as we showed separately) to abort the drink.

But how should we measure predictability? The graph below shows one way. It represents a single drink in the same way as before. But each point on the graph of eye position has arrows attached to it. The length of the arrow signifies, for each frame of film, the likelihood (as totted up over all the drinks by all the chicks) that the eye height in the next frame will be lower, higher, or the same.

You can see that during the upstroke, when the bird is allowing the water to trickle down its throat, there is a high probability that the upstroke will continue its graceful curve in the upward

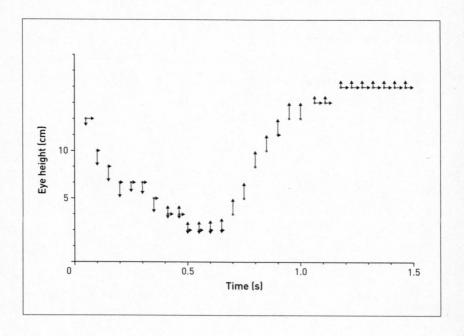

direction. A decision to perform an FAP is being carried out, with no further decisions during its course. But during the downstroke there is more unpredictability. For each frame of the downstroke, the height of the eye in the next frame is undecided between lower or the same, and there is even some likelihood that it might be higher – that is, that the drink will be aborted.

Could we use these arrows to compute an index of uncertainty or 'decisioniness'? The index we chose was based on information theory, devised in the 1940s by the inventive American engineer Claude Shannon. The information content of a message can be informally defined as its 'surprise value'. Surprise value is a convenient opposite of predictability. Classic examples are 'It is raining in England today' (low information content because no surprise) versus 'It is raining in the Sahara Desert' (high information content because surprising). For reasons of mathematical convenience, Shannon computed his index of information content in *bits* (short for 'binary digits'), by summing up the logarithm (base 2) of the prior probabilities that were open to doubt before the message was received. The information content of a penny toss is one bit, because the prior uncertainty is heads or tails – two equiprobable alternatives. The information content of a playing card's suit is two bits (there are four equiprobable alternatives and the base two logarithm of four is two, corresponding to the minimum number of yes/no questions you'd need to ask in order to establish the suit). Most real examples are not so simple, and the possible outcomes are usually not equiprobable, but the principle is the same and a version of the same mathematical formula conveniently does the trick. It was this mathematical convenience that led us to use the Shannon Information Index as our measure of predictability or uncertainty.

Once again we have a graph (overleaf) of eye height against time during a drink. The thin lines represent times of low predictability, or high probability of a decision intervening to change the future.

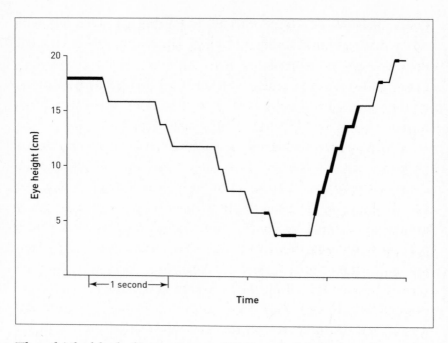

The thick black lines represent times of high predictability (information content less than an arbitrary threshold of 0.4 bits), during which a decision is being carried out and no new decision is expected. The upstroke is predictable once it starts, but the down-stroke is not. The pause between drinks is predictable for the rather boring reason that the pause is most likely to continue into the next frame – it's hard to predict when the next drink will start.

As always, keep in mind that the particular behaviour, here drinking, is not of interest in itself. Drinking in chicks was a stand-in for behaviour generally, just as pecking was in my doctoral research. We were interested in the very idea of a decision and – in the case of drinking – whether we could identify moments of deci-sion. We were trying to explore a way to demonstrate the very existence of a Fixed Action Pattern, rather than simply taking it for granted as ethologists were wont to do.

We adopted a different approach in our next project on decision-making, which was a study of self-grooming in flies. Ethologists often ask whether, if you know what an animal is doing now, you can predict what it will do next. Marian and I wanted to know whether, sometimes, you can predict what it will do in the more distant future *better* than you can predict what it will do in the more immediate future. This might be true, for instance, if behaviour is organized like human language. There are times when the beginning of a sentence predicts how the sentence will end better than it predicts the middle – which might contain any number of embedded adjectival or relative clauses, for instance. 'The girl hit the ball' is a sentence whose beginning demands something like the ending, whether or not there are embedded adjectives or adverbs or clauses in the middle: '**The girl** with red hair, who lives next door, vigorously **hit the ball.**'

We didn't find evidence of language-like grammatical structure in the grooming behaviour of flies (although see below). But we did find an interesting zigzag pattern in the way predictability decays over time: in other words, the immediate future may be less predictable than the (slightly) more distant future. I'll outline our research briefly here and not in detail, because it's a bit complicated.

Flies are not normally seen as beautiful, but the way they wash their faces and their feet is rather dear. Look next time a fly lands on you: you'll very probably see the behaviour. It may rub its front feet together, or wipe its great big eyes with them. It may rub the middle foot on one side against the hind foot on the same side, or clean its abdomen or wings with its hind feet. Somewhere inside that tiny head, decisions are spontaneously being generated, and a fair number of those decisions concern which bit of the body to clean next. The appeal of self-grooming behaviour for us was that the fly's choice of behaviour was unlikely to be externally stimulated. We presumed that external stimulation amounted to an ever-present need to keep clean – ever-present in the sense that, though important, it was unlikely to determine exactly when a particular grooming action would be chosen. Dirty wings would impair flight. Dirt would impair the highly sensitive tasting organs in the feet, which flies use to decide whether or not to stick out the tongue and eat. So cleaning is important. But presumably the decision about which bit to clean is not determined by the sudden arrival of a new piece of dirt. Rather, we suspected that these rapid, moment-to-moment decisions were internally generated by unseen fluctuations deep inside the nervous system.

We recognized eight distinct grooming acts, which we presumed would show up as FAPs if we had the time to do a frame-by-frame analysis like the one we did with the chicks drinking: FR (rub front feet together), TG (rub tongue between front feet), HD (wipe head with front feet), FM (rub one or other middle foot between front feet), BM (rub one or other middle foot between back feet), BF (rub back feet together), AB (wipe abdomen with back feet), WG (wipe wings with back feet). Using a Dawkins Organ, we recorded the sequences of these eight grooming acts, plus MV (move away) and NO (stand still, doing nothing).

The graph below shows the probability, given that the fly is now doing HD, that it will do FR next ('lag' = 1, probability very high), next but one (very low probability), next but two (high probability), next but three (low probability) etc. You can see that there is a

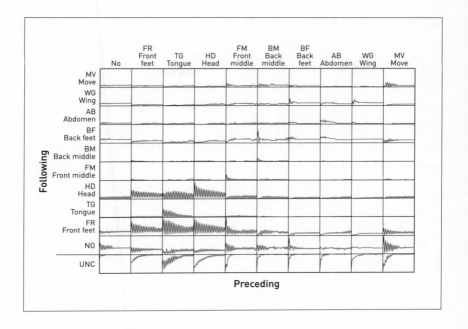

pronounced tendency to alternate, and also that there is a general die-away (as you'd expect) in predictability as we look to the more distant future – longer and longer 'lags'.

That picture was for the particular case of FR following HD. We plotted the same kind of graph for all the possible transitions, and put the graphs into a table (above).

You can see that many of the transitions follow the same zig-zag pattern, although some are exactly out of phase with each other. The bottom row (UNC) shows the uncertainty attached to predictions of the future, following each behaviour, calculated using the Shannon Information Index, in the same way as for the chick drinking study.

We also tried the experiment of using the human ear to identify patterns in animal behaviour. For this, we used a Dawkins Organ rendition of fly grooming behaviour, but eliminated the true intervals between musical notes. I told the computer to reduce all the intervals to a single standard short interval and then we simply listened to the 'music'. It sounded rather like 'modern' (as opposed to 'traditional') jazz. Also rather like the 'singing' Elliott computer of my juvenile insomniac dalliance – I suppose the comparison might be interesting. I thought the human ear might be a promising piece of apparatus to use in detecting patterns in animal behaviour, but didn't follow up the method; I merely report it here as an interesting curiosity. If the World Wide Web had existed in those days, I would surely have uploaded the fly washing music and you could now dance to it. As things are, I'm afraid those Dipteran Melodies are gone for ever, like the Lost Chord.

I cannot claim that our fly study, or the other studies of decision-making that preceded it, really tell us much about how animal brains work. I see them more as explorations of methods: not just methods of doing research on animal behaviour, but methods of *thinking*. Marian and I did a lot more work on the flies, but it's all published and I won't write any more about it here. It did, however, feed into my next big writing project: a long theoretical paper on 'Hierarchical organisation as a candidate principle for ethology'. This is the subject of a later section.

Meanwhile, in 1973, Niko Tinbergen won the Nobel Prize in Physiology or Medicine (jointly with Konrad Lorenz, his co-founder of ethology, and Karl von Frisch, the discoverer of the legendary bee dance). Just one year later, in 1974, Niko reached Oxford's mandatory retirement age of sixty-seven, and the university agreed to appoint a successor as Reader in Animal Behaviour. 'Reader' was a rather prestigious rank at Oxford, now, I think, fallen into desuetude in a move to bring the title of 'professor' into line with American custom by sprinkling it about more liberally – the rather unkindly dubbed 'Mickey Mouse professors'. I was very content where I was as lecturer, and had no ambition to apply for the job.

Most people thought of Mike Cullen as Niko's natural successor. Perhaps for that very reason, in order to make a clean break, the majority of the appointment committee went for David McFarland. As Hans Kruuk wrote in his biography of Tinbergen, 'one could hardly have found anyone more unlike Niko'. Though controversial in many quarters, David's appointment was in some ways an inspired one, at least if you take the view that a new appointment is an opportunity for a new departure. His science was highly theoretical, indeed mathematical. He brought to it the intuitions of a mathematician, and he surrounded himself with trained mathematicians and engineers who could do the algebra. The talk in the coffee room switched from gulls and sticklebacks in the field to feedback control systems and computer simulations.

Perhaps it was a microcosm of the way biology was changing. I was young and not yet set in my ways. 'If you can't beat them, join them' was my attitude. So I set to work to learn control theory from the engineers and mathematicians who now surrounded me. And what better way to learn it than hands-on? I again indulged my passion – or vice – for computer programming, and wrote a program for a digital computer ('my' PDP-8), enabling it to behave like an analogue computer. To this end, I invented yet another computer language, which I called SysGen.

Unlike the propositions in a conventional computer language like Fortran, which are executed sequentially, SysGen statements were executed 'simultaneously' – not *really* simultaneously, of course, because a digital computer does everything sequentially at bottom; but they could be written in any order. My task in writing the SysGen Interpreter program was to persuade the digital computer to behave as *if* the operations were simultaneous: a virtual analogue computer. As with an analogue computer, results were displayed as a set of graphs on an oscilloscope screen.

I'm not sure how useful SysGen was in practice, but inventing the language, and writing the Interpreter program for it, certainly helped me to understand not just control theory but also the integral calculus. It gave me a much better idea of what it means to integrate. I was mindful of my maternal grandfather's recommendation of *Calculus Made Easy*, by his old mentor Silvanus Thompson (who, as quoted earlier, was fond of saying, 'What one fool can do, another can'). Thompson introduces his explanation of integration with another phrase that has stuck in my brain: 'So we had best lose no time in learning how to integrate.' I had only half understood integration in Ernie Dow's lessons, and SysGen gave me the sort of hands-on experience that assists comprehension.

Similar in intention, but much easier and less time-consuming, was my attempt to understand Chomsky-style linguistics by the hands-on method. I wrote a computer program to generate random sentences, which might not have been very meaningful but were always scrupulously grammatical. This is easy – and that very fact is instructive – given that your programming language allows procedures (subroutines) to call themselves *recursively*. This was true of Algol-60, the programming language that I favoured at the time under the influence of Roger Abbott, who had brilliantly succeeded in writing an Algol compiler for the PDP-8. Algol subroutines could call themselves, unlike the contemporary version of that old workhorse of scientific programmers, IBM's Fortran language. Mention of Fortran reminds me of a nice in-joke told by Terry Winograd, pioneer of artificial intelligence. Some time in the 1970s I attended a fascinating conference in Cambridge on the state of the art of artificial intelligence programming, and Winograd was the star lecturer. At one point he gave vent to a wonderful piece of sarcasm: 'Now, you may be one of those who says, "Fortran was good enough for my *grandfather*, it's good enough for me."'

Given that your programming language allows procedures to call themselves recursively, writing programs to deliver correct grammar is remarkably – elegantly – easy. I wrote a program that had procedures with names like NounPhrase, AdjectivalPhrase,

PrepositionalClause, RelativeClause etc., all of which could call any other procedure, including themselves, and it generated random sentences like this one:

(The adjective noun (of the adjective noun (which adverbly adverbly verbed (in noun (of the noun (which verbed))))) adverbly verbed)

Parse it carefully (as I have done here using brackets, although the computer didn't generate them but left them implicit) and you'll see that it is grammatically correct although not exactly dripping with information. It makes syntactic but not semantic sense. The computer could easily inject semantics (if not sense) by replacing 'noun', 'adjective' etc. with particular, randomly chosen instances of nouns and adjectives. Thus you could inject a vocabulary from a chosen domain, such as pornography or ornithology. Or you could inject the vocabulary of francophoney metatwaddle – as Andrew Bulhak was later to do when he wrote his hilarious 'Postmodernism Generator', which I quoted in *A Devil's Chaplain*:

If one examines capitalist theory, one is faced with a choice: either reject neotextual materialism or conclude that society has objective value. If dialectic desituationism holds, we have to choose between Habermasian discourse and the subtextual paradigm of context. It could be said that the subject is contextualised into a textual nationalism that includes truth as a reality. In a sense, the premise of the subtextual paradigm of context states that reality comes from the collective unconscious.

This randomly generated garbage makes about as much sense as many a journal devoted to the metatwaddle of 'literary theory', and Bulhak's program is capable of generating a literally indefinite quantity of it.

Two more programming projects date from around this time in my life, both of which also, as it turned out, served to hone my skills for the future rather than deliver results of more immediately practical usefulness. The first of these was a program to translate from one computer language to another: specifically, from BASIC to Algol-60. It worked well for those two languages, and would have worked, with minor detailed changes, to translate from any computer language of that general algorithmic type to any other. My second project of this time was STRIDUL-8: a program to make the PDP-8 computer sing like a cricket.

I had been inspired to work on crickets by my Berkeley friend, the neurobiologist David Bentley; and my entomologically inclined graduate student Ted Burk (now a professor in Nebraska) was keen to do his doctoral thesis on them. David kindly sent me some eggs of the Pacific field cricket *Teleogryllus oceanicus*. They hatched in Oxford and soon we had a thriving colony, which Ted looked after, feeding them on lettuce. While Ted productively pursued his own research on the crickets' behaviour, I conceived a parallel project using computer-generated courtship song. That research project was never completed, but I did complete the writing of STRIDUL-8, and it worked pretty well.

My testing apparatus was a seesaw, made out of balsa wood so it was very light – as it had to be for crickets. It was really just a long balsa-wood passage, closed with netting at each end and over its top, and resting on a hinged fulcrum in the middle. Only one cricket was placed in the passage at a time, and it was free to walk from one end to the other as often as it liked. Whichever end it approached tipped down, as a seesaw should, and this fact was recorded by a micro-switch, which importantly also reversed the location of the sound. There were two little loudspeakers, one at each end of the seesaw. Cricket song was played through whichever of the two loudspeakers was at the opposite end of the seesaw from the cricket.

So, imagine you are a female cricket, sitting somewhere towards the west end of the corridor. Song is playing from the east. You like what you hear, so you start to walk east. When you near the east end, your weight tilts the seesaw to the east, tripping the micro-switch and thereby informing the computer, which now switches the song to the west end loudspeaker. So you turn and walk west, and the whole process happens in reverse. Preferred songs therefore generate a large number of seesaw reversals and these were automatically counted by the computer. Whether the cricket thought she was pursuing an ever-receding coy male, or whether she thought the male was jumping capriciously over her head, or whether she thought at all, is impossible to say. Unpreferred songs would generate only a small number of rockings of the seesaw. Indeed, if a song was positively aversive, the cricket would stay at the opposite end of the passage and not generate any seesaw tips at all.

That, then, was my apparatus for measuring how much crickets like songs of different types. Play Song *A* for five minutes of the alternating seesawing regime, then do the same with Song *B*, and so on for many trials, properly randomized etc. Count seesaw tippings as a measure of how much the cricket liked each song. The point about computer-generated songs, as opposed to real ones, was to try to dissect out, in classic Tinbergen fashion, what it is about their own species song that crickets like. The computer would vary how it sang, in systematic ways. The initial plan was to start with a simulation of the species' natural song and then change it – drop bits out, enhance other bits, vary the interval between chirps and so on. Later – it was a somewhat wild hope – I envisaged the computer being programmed instead to start with random song and 'learn' – or we could equivalently say 'evolve' – choosing 'mutations' step by step, until it progressively homed in on a synthetic preferred song. If the preferred song had turned out to be the natural song of *Teleogryllus oceanicus*, wouldn't that have been sensational? And then if I had done the same thing with *Teleogryllus commodus* and the computer had homed in on its rather different song. What bliss that would have been for the researcher!

In programming the computer to sing, I wanted to make it as versatile as possible. Versatility is what computers are good at. As with the analogue computer simulation, as with the language translation program, I wanted to program the general case. And this is where STRIDUL-8 came in: its language allowed you to specify any combination of pulses and intervals, and therefore any cricket song in the world. STRIDUL-8 had an intuitively reasonable bracket notation, which enabled the user to insert repeats, and repeats embedded within repeats, in a manner reminiscent of the grammar of language (see pages 229 and 235).

STRIDUL-8 worked well. Its simulations of cricket song sounded like real crickets to human ears, and it was easy to program the computer to sing like any cricket species in the world. However, when I demonstrated the system to Dr Henry Bennet-Clark, world authority on the acoustics of insect sounds, newly arrived from Edinburgh to take up a position in Oxford, he made a face and said 'Eeugh!' STRIDUL-8 could only specify the pattern in time of pulses of sound, each pulse corresponding to one stroke of the wings against each other. I had made no attempt to simulate the actual wave form produced by each wing stroke, and this is what Henry objected to. He was right. STRIDUL-8, as it stood, could not have done justice to the European tree crickets of whose song Henry once wrote that if moonlight could be heard that is how it would sound. Temporarily discouraged, I put my whole cricket song project on the back burner, while I attended to other pressing tasks, notably a challenging invitation from Cambridge. And unfortunately I never returned to it: my cricketing days were over. I've often regretted it. I think most scientists have sad loose ends, projects started, never finished. If I ever had vague intentions to return to the crickets, they were thwarted by Moore's Law: computers change so rapidly that if you leave a loose end of research untied for as long as I left mine, you find that the extant computers have all become newer, sexier models, and they have forgotten how to run your earlier programs. To find a computer that would run STRIDUL-8 today I'd have to go to a museum.

THE GRAMMAR OF
BEHAVIOUR

THE Oxford Animal Behaviour Research Group under Tinbergen had long maintained cordial relations with the corresponding sub-department at Cambridge, housed in the neighbouring village of Madingley. 'Madingley' was founded in 1950 by W. H. Thorpe – a distinguished scientist whose gently austere, almost ecclesiastical personality is best summed up by Mike Cullen's jest that it was entirely appropriate that when Thorpe needed a notation for recording birdsong, he transcribed it for the *organ*. Madingley celebrated its quarter-century in 1975 with a conference in Cambridge organized by Patrick Bateson and Robert Hinde, the leading figures of the Madingley group after Thorpe's retirement, both of whom later became heads of Cambridge colleges. Many of the speakers at the Madingley conference were past or present members of that group, but they invited some outsiders too, and David McFarland and I were honoured to be the Oxford contingent.

Nowadays, on the rare occasions when I agree to speak at such a conference, I confess that I usually find myself dusting off a previous talk and updating it. Younger and more vigorous in 1974, I took the risk of pushing the boat out and undertaking to write something entirely new for Madingley's jubilee conference and the book that came out of it. The topic I chose, 'hierarchical organization', had a track record in the history of ethology. It was the main theme of one of the boldest – and most criticized – chapters in Tinbergen's magnum opus *The Study of Instinct*, the chapter entitled 'An attempt at a synthesis'. I took a rather different approach – or, rather, several different approaches, and I too attempted a synthesis.

The essence of hierarchical organization, as I interpreted it, is the idea of 'nested embedment'. I can explain this by contrast with what it is *not*, and this is where I echo the discussion above about grammar. You might attempt to describe the stream of events – the stream of things an animal does, say – as a Markov Chain. What's that? I won't attempt a formal, mathematical definition such as was offered by the Russian mathematician Andrey Markov. An informal, verbal definition is this. A Markov Chain of animal behaviour is a series in which what an animal does now is determined by what it did previously, back a fixed number of steps but no further. In a first-order Markov Chain, what the animal does next can be predicted statistically from its immediately preceding action, and not from anything earlier. Looking at the last but one action (last but two action etc.) gives you no additional predictive power. In a second-order Markov Chain, you improve your ability to predict if you look at the previous two actions, but no further back than that. And so on.

Hierarchically organized behaviour would be very different. Markov Chain analysis, of any order, wouldn't work. The predictability of behaviour wouldn't decay smoothly as you look into the future, but would jump up and down in an interesting way – like with the blowfly grooming, but more interestingly than that. In an ideal case, behaviour would be organized in discrete chunks. And chunks within chunks. And chunks within chunks within chunks. That's what's meant by nested embedment. The clearest model for nested embedment is syntax, the grammar of human language. Think back to the program that I wrote to generate random grammatical sentences, and the example sentence that I quoted:

> **The adjective noun** of the adjective noun which adverbly adverbly verbed in noun of the noun which verbed **adverbly verbed**.

The core sentence is in bold. You can read it and it is grammatically correct, without the various embedded relative clauses or prepositional clauses in the middle. We can build up the embedment as follows. The important point is that the build-up can occur *inside* the core sentence, or inside already embedded parts of the sentence. Read to yourself the emboldened parts of the following:

> **The adjective noun** of the adjective noun which adverbly adverbly verbed in noun of the noun which verbed **adverbly verbed**.

> **The adjective noun of the adjective noun** which adverbly adverbly verbed in noun of the noun which verbed **adverbly verbed**.

> **The adjective noun of the adjective noun which adverbly adverbly verbed** in noun of the noun which verbed **adverbly verbed**.

> **The adjective noun of the adjective noun which adverbly adverbly verbed in noun** of the noun which verbed **adverbly verbed**.

> **The adjective noun of the adjective noun which adverbly adverbly verbed in noun of the noun which verbed adverbly verbed**.

In every member of the sequence above, you can read the emboldened part on its own, and discover that it is grammatically correct. You can delete the unbold, embedded bits and it might change the meaning but it doesn't stop the sentence from being grammatically correct.

If, on the contrary, you were to build up the sentence by adding the bits progressively from left to right, none of the series would be grammatical until you hit the end of the whole sentence.

The adjective noun [not a sentence]

The adjective noun of the adjective noun [not a sentence]

The adjective noun of the adjective noun which adverbly adverbly verbed [not a sentence]

The adjective noun of the adjective noun which adverbly adverbly verbed in noun [not a sentence]

The adjective noun of the adjective noun which adverbly adverbly verbed in noun of the noun which verbed adverbly verbed [finally, we have a sentence].

Only in the very last case does the sentence achieve closure and become grammatical. What I wanted to know was whether animal behaviour is organized as a Markov Chain, or whether it is organized in a nested embedded way, perhaps like syntax or perhaps in some other way hierarchically embedded. You can see that there were inklings of the idea lurking behind the research that Marian and I did on drinking in chicks and especially self-grooming in flies. Now, in my Madingley paper, I wanted to look more generally at the question of hierarchical organization, from a theoretical point of view as well as by looking at real studies of animal behaviour.

After defining various kinds of hierarchy in a convenient notation of mathematical logic, I considered possible evolutionary

This picture of me with my parents, taken at a family wedding (my sister Sarah was a bridesmaid, so not with us at this point), unfortunately does not show the bright red of the cap I wore as a Chafyn Grove schoolboy. In my first term at Oundle I don't think I was as happy as I looked for the camera. One of the best things about that school was Ioan Thomas, caught here encouraging an appetite for wonder at the living world.

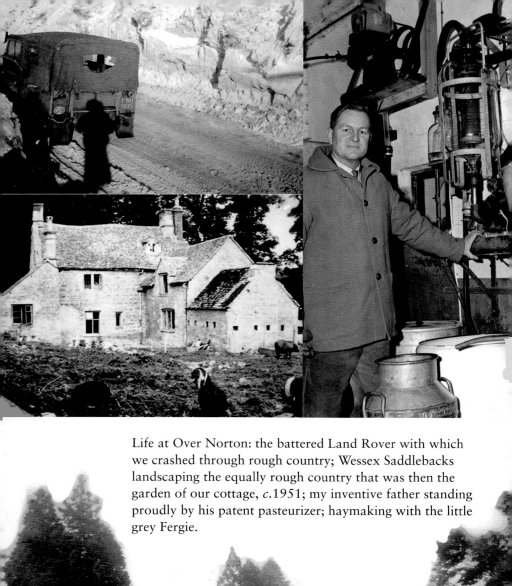

Life at Over Norton: the battered Land Rover with which we crashed through rough country; Wessex Saddlebacks landscaping the equally rough country that was then the garden of our cottage, c.1951; my inventive father standing proudly by his patent pasteurizer; haymaking with the little grey Fergie.

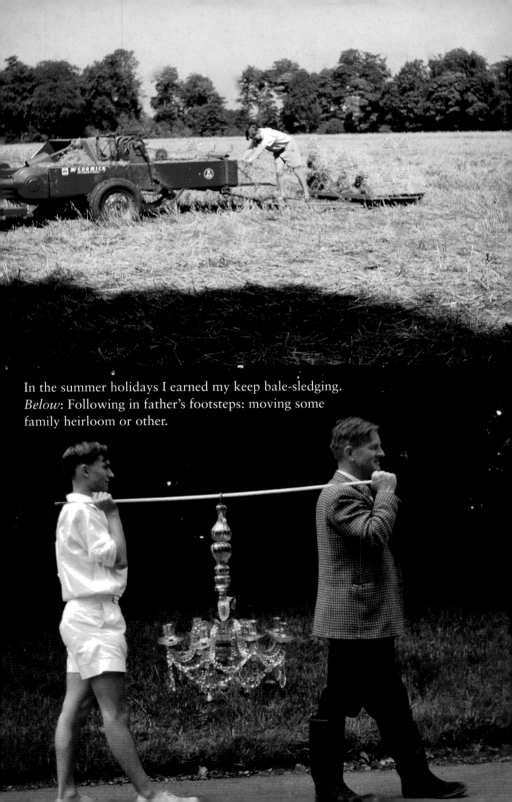

In the summer holidays I earned my keep bale-sledging.
Below: Following in father's footsteps: moving some
family heirloom or other.

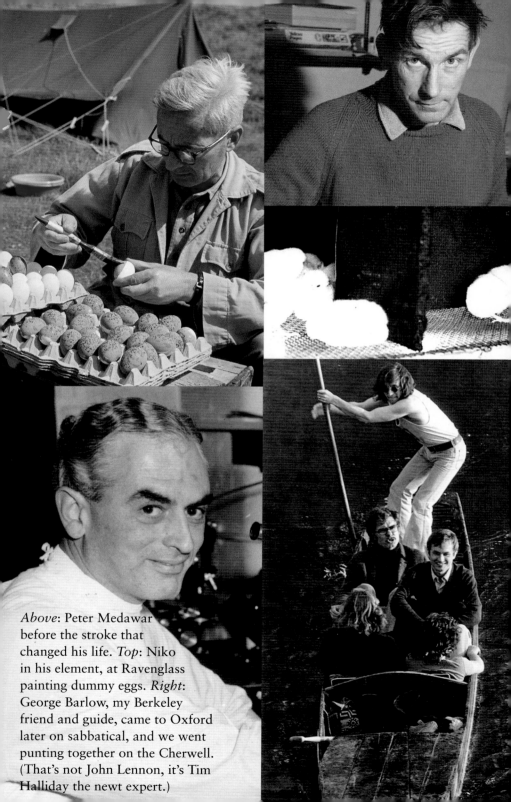

Above: Peter Medawar before the stroke that changed his life. *Top*: Niko in his element, at Ravenglass painting dummy eggs. *Right*: George Barlow, my Berkeley friend and guide, came to Oxford later on sabbatical, and we went punting together on the Cherwell. (That's not John Lennon, it's Tim Halliday the newt expert.)

Left, top: 'The deeply intelligent eyes, understanding what you meant even before the words came out . . . sceptical, quizzical tilt of the eyebrows, under the untidy hair.' Mike Cullen, sadly missed mentor to so many. *Left*, middle: What to peck? Chicks that had never seen overhead light. *Above*: Hunting the Surrey puma; intrepid explorer scours the landscape for wild beasts. *Below*: Wild beasts or frightened boys? California National Guard raggedly confront the Peace People in Berkeley.

Top left: Cricket commentary: Ted Burk and I recording behaviour with microphone and Dawkins Organ. *Middle left*: The Animal Behaviour Research Group after the move from Bevington Road. Marian is far left. I am slightly right of centre. *Bottom left*: A PDP-8 computer like the one that fed my addiction in 13 Bevington Road. *Bottom right*: Professor Pringle and (left to right) his colleagues, E. B. Ford, Niko Tinbergen, William Holmes, Peter Brunet, David Nichols. *Left*: Danny Lehrman (standing) and Niko Tinbergen (right) settling their differences. *Below*: Niko in his element again: will the ash fall from his cigarette before he finishes the shot?

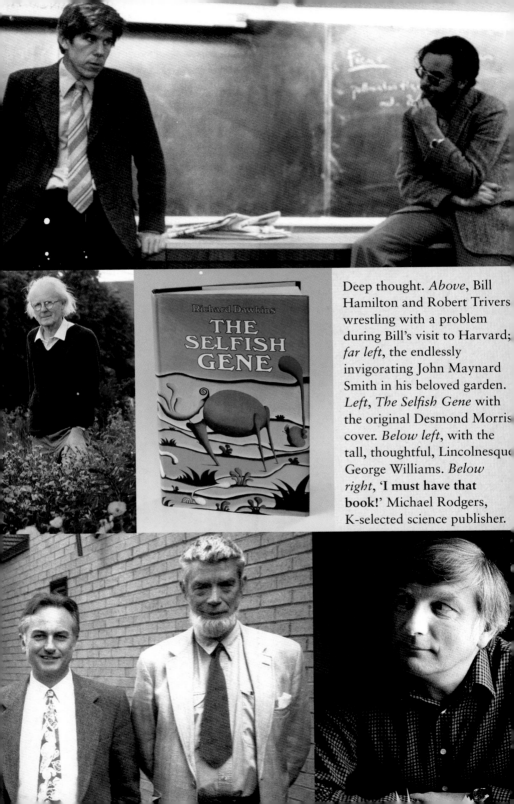

Deep thought. *Above*, Bill Hamilton and Robert Trivers wrestling with a problem during Bill's visit to Harvard; *far left*, the endlessly invigorating John Maynard Smith in his beloved garden. *Left*, *The Selfish Gene* with the original Desmond Morris cover. *Below left*, with the tall, thoughtful, Lincolnesque George Williams. *Below right*, 'I must have that book!' Michael Rodgers, K-selected science publisher.

advantages of hierarchical organization. To illustrate what I called the 'evolutionary rate advantage' I borrowed from the Nobel Prize-winning economist Herbert Simon a parable of two watchmakers called Tempus and Hora. Their watches kept equally good time, but Tempus took much, much longer to complete a watch. Both kinds of watch had 1,000 components. Hora, the more efficient watchmaker, worked in a hierarchical, modular way. He put his components into 100 sub-assemblies of ten components each. These in turn were assembled into ten larger units, which were finally put together to complete the watch. Tempus, on the other hand, tried to put all 1,000 components together in a single large assembly operation. If he dropped one component, or if the telephone interrupted him, the whole caboodle fell to bits and he had to start again. He very rarely completed a watch, while Hora, with his hierarchical modular technique, was churning them out. The principle will be familiar to all computer programmers, and it surely applies to evolution and to the building of biological systems.

I also extolled another advantage of hierarchical organization, the 'local administration advantage'. If you are trying to control an empire from London, or in earlier times from Rome, you cannot micro-manage what happens in remote parts of the empire, because communication channels – in both directions – are too slow. Instead, you appoint local governors, give them broad policy directives, and leave them to take day-to-day decisions on their own. The same necessarily applies to a robot vehicle on Mars. Radio signals take several minutes to travel the distance. If the vehicle encounters a local difficulty, say a boulder, it sends the information back to Earth, and it takes four minutes to get there. 'Turn left to avoid boulder,' flashes back the urgent reply, and that again takes four minutes to reach Mars. Meanwhile, the wretched vehicle has long since ploughed into the boulder. Obviously the solution is to delegate local control to an on-board computer, and give the local computer only general policy instructions like: 'Explore the crater to the north-west, taking care to avoid boulders whenever you encounter them.' By the same token, if there are several vehicles exploring different parts of Mars, it makes sense for Earth to send general policy instructions to one senior computer on Mars, which sends more detailed instructions coordinating the activities of all its subordinate vehicles, each with its own on-board computer to take fine-grained local decisions. Armies and business corporations use similar hierarchical chains of command, and once again biological systems do the same.

Especially pleasing in this connection are the giant dinosaurs whose very long spinal cord imposed an inconvenient distance between the brain in the head and the seat of much of the action, the giant hind legs. Natural selection solved the problem with a second 'brain' (enlarged ganglion) in the pelvis:

> Behold the mighty dinosaur,
> Famous in prehistoric lore,
> Not only for his power and strength
> But for his intellectual length.
> You will observe by these remains
> The creature had two sets of brains—
> One in his head (the usual place),
> The other at his spinal base.
> Thus he could reason 'A Priori'
> As well as 'A Posteriori'.
> No problem bothered him a bit
> He made a head and tail of it.
> So wise was he, so wise and solemn,
> Each thought filled just one spinal column.
> If one brain found the pressure strong
> It passed a few ideas along.
> If something slipped his forward mind

'Twas rescued by the one behind
And if in error he was caught
He had a saving afterthought.
As he thought twice before he spoke
He had no judgment to revoke.
Thus he could think without congestion
Upon both sides of every question.
Oh, gaze upon this model beast,
Defunct ten million years at least.

Bert Leston Taylor (1866–1921)

'Thus he could reason "*A Priori*" / As well as "*A Posteriori*"' – I wish I'd written that. You'd have to look far before you found another poem with quite so many flashes of clever wit in almost every line.

Having established the advantages of hierarchical organization more generally, I moved on to see whether there was evidence of it in specific cases of animal behaviour. Beginning by re-analysing the data Marian and I had recorded from blowflies, I moved on to other data from the animal behaviour literature, which I ferreted out in the library. Among others, I included a large study on the behaviour of damsel fish, another on face-grooming behaviour by mice and another on the courtship behaviour of guppies.

I wanted to devise mathematical techniques for detecting hierarchical embedment, in an attempt at objectivity, unbiased by my own preconceptions. Here's just one of several computer-based methods I thought up. This one I dubbed Mutual Replaceability Cluster Analysis. My method started by counting frequencies of transitions between behaviour patterns, but then analysed the data in a special hierarchical way. I fed into the computer a table showing how many times each behaviour pattern in the animal's repertoire was followed by each other one. Then the computer systematically examined the data to see if it could find pairs of behaviour patterns

that were *mutually replaceable*. Mutually replaceable means that you could stick either of them in the place of the other and the overall pattern of transition frequencies would remain the same (or near enough the same, according to some previously defined criterion). Once a mutually replaceable pair had been identified, both members of the pair were renamed with a *joint* name, and the table of transitions contracted because it now had one fewer rows and one fewer columns. Then the contracted table was fed back into the cluster-analysis program, and the whole thing was repeated as many times as necessary to use up the whole list of behaviour patterns. As each pair of behaviour patterns was swallowed up in a cluster, or as each already swallowed cluster was swallowed up in a bigger cluster, the program moved up one node in a hierarchical tree. Opposite, for instance, is my Mutual Replaceability tree for the behaviour patterns of guppies, using data from a group of Dutch workers led by Professor G. P. Baerends (who, incidentally, had been Niko Tinbergen's first graduate student and later became one of the leading figures in European ethology).

The upper diagram shows the transition frequencies of guppy behaviour patterns, as measured by the Dutch scientists. Each circle is labelled with the code name of a behaviour pattern, and the thickness of the lines shows the frequency of transition from one to the other (black lines move from left to right, grey lines from right to left). The lower diagram shows the results of feeding the same data into my Mutual Replaceability Cluster Analysis program. The numbers represent the numerical index of mutual replaceability that I used to compare with the criterion for deciding to unite two entities (actually a rank correlation coefficient, if you happen to be interested). I got similar hierarchical trees for the damsel fish, the mice, Marian's and my blowflies etc.

Yet another way of thinking about hierarchy, which I used in my Madingley paper, is the hierarchy of *goals*. A goal is not necessarily a consciously held goal in the animal's brain (although it might be). I simply meant a condition which brings behaviour to an end. For example, complicated sequences of prey-catching behaviour in a cheetah would be brought to a close by the 'goal state' of a successful kill. But goals can be hierarchically embedded within each other, and that is a fruitful way to look at it. I made a distinction between 'action rules' and 'stopping rules'. An action rule tells the animal (or computer in the case of a computer simulation) exactly what to do and when to do it, including lots of conditional instructions (IF . . . THEN . . . ELSE etc). A stopping rule tells the animal (or computer simulation): 'Behave at random (or try out lots of possibilities) and don't stop until the following *goal state* is achieved' – say, a full stomach.

A pure action-rule program for a complicated task like hunting by a cheetah would become impossibly elaborate. Much better to use stopping rules. But not just one big stopping rule – behave at random until the goal state of full stomach is achieved. Any cheetah living by that rule would die of old age before achieving a square meal! Instead, the sensible way for natural selection to have programmed the behaviour would be with hierarchically embedded stopping rules. The global goal (continue until stomach is full) would 'call up' subsidiary goals such as 'walk around until gazelle sighted'. The goal state 'gazelle sighted' would terminate that particular stopping rule and initiate the next one: 'drop down and crawl slowly towards gazelle'. That would be terminated by the goal state 'gazelle now within striking distance'. And so on. Each of these subsidiary stopping rules would call up its own, internally embedded stopping rules, each with its own goal state. At much lower levels, even individual muscle contractions often conform to the design that engineers call 'servo-control'. The nervous system specifies a target state for a muscle, which contracts until the target state ('stopping rule') is achieved.

But I earlier introduced the idea of hierarchical embedment by using the analogy of human grammar. My Madingley paper finally returned to this fascinating topic, and asked whether there was any evidence that animal behaviour had something equivalent to grammatical structure. If it did, this would be extremely interesting, because it might give us some inklings of the evolutionary antecedents of human language. When true language, with true hierarchical syntax, finally evolved in humans, dare we speculate that it was able to build on a ready-made foundation of pre-existing neural structures that were put in place for different reasons, nothing to do with language, long ago?

The earliest attempt to look at this question was made by my Oxford colleague John Marshall, a linguist. He used courtship behaviour of male pigeons, taking data from the published etho-logical literature. There were seven 'words' of the pigeon lexicon: things like Bow (to the female), Copulate, etc. Marshall used his skills as a linguist to postulate a 'phrase structure grammar', just as Chomsky had before him for human language. For my Madingley paper, I translated Marshall's grammar into the (now largely obsolete) computer language that I favoured at the time, Algol-60. Readers familiar with computer programming will note that, once again, the program is heavily recursive – procedures call themselves, the very essence of hierarchical embedment as I have already explained. In the program, 'p' was replaced by 'If some probability condition, such as 0.3, is met . . .'

At the top of the diagram overleaf is Marshall's 'phrase structure grammar' for pigeon courtship behaviour. In the middle is my Algol-60 translation. And at the bottom are several sequences of 'behaviour' generated by my program.

Unfortunately, Marshall's analysis doesn't really allow us to draw any secure conclusion about pigeons. How do we know whether the grammar he proposed is 'correct'? In the case of human syntax, any native speaker of the language can immediately tell you if it is

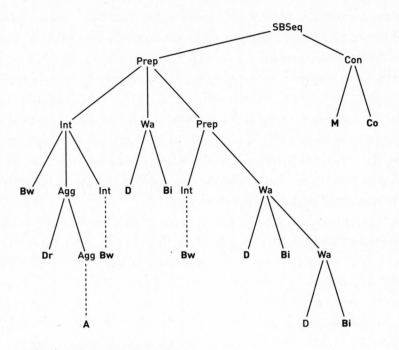

begin comment Marshall's pigeon grammar;
procedure SBSeq; **begin** Prep; Con **end**;
procedure Prep; **begin** Int; Wa; **if** p **then** Prep **end**;
procedure Int; **begin** "BW"; **if** p **then** Agg; **if** p **then** Int **end**;
procedure Agg; **begin if** p **then** "DR"; **if** p **then** "A"; **if** p **then** Agg
         end;
procedure Wa; **begin** "D"; "**BI**"; **if** p **then** Wa **end**;
procedure Con; **begin** "M"; "CO" **end**;
Boolean procedure p;
      begin comment true or **false** at random. Probability manipulated.
      end;
start: SBSeq; **goto** start
end of pigeon grammar;

Sample results of running the programme:
BW DR D M CO
BW A D BI BW DR D BW A D BW A D BI M CO
BW A D BI M CO
BW DR D BW DR D BI BW DR D BI BW A D BW A D M CO

correct. Marshall had no such checking mechanism. As with much of the research that I did during this period, my goal state was not so much to find something permanently true about particular animals as to find novel and exciting ways in which animal behaviour might be studied in the future.

The Madingley paper[1] represented a kind of closure for me, a climax to the first part of my scientific career, beginning in my early twenties and ending in my early thirties. At this point, I took off in an entirely new direction, never to return to those youthful mathematical pastures. That new direction, which was to define the rest of my career, and approximately the second half of my life, opened up with the publication of my first book, *The Selfish Gene*.

[1] R. Dawkins, 'Hierarchical organization: a candidate principle for ethology', in P. P. G. Bateson and R. A. Hinde, eds, *Growing Points in Ethology* (Cambridge, Cambridge University Press, 1976), pp. 7–54.

THE IMMORTAL GENE

IN 1973, strike action by the National Union of Mineworkers led to a crisis in which the Conservative government of Edward Heath imposed a so-called 'three-day week' in Britain. In order to conserve fuel stocks, electricity for non-essential purposes was rationed. We were limited to three days per week, and there were frequent power cuts. My cricket research depended on electricity but writing didn't – in those days I wrote with a portable typewriter on, of all outlandish surfaces, sheets of flat white stuff called paper. So I decided to call a temporary halt to my cricket research and begin work on my first book. This was the genesis of *The Selfish Gene*.

Selfishness and altruism and the whole idea of a 'social contract' were much in the air at the time. Those of us on the political left tried to balance our sympathy for the miners on the one hand with hostility to what some saw as their strong-arm tactics, holding society at large to ransom, on the other. Did evolutionary theory have anything to contribute to this important dilemma? The previous decade had seen a string of popular science books and television documentaries, gallantly attempting to apply Darwinian theory to questions of altruism and selfishness, collective versus individual welfare, but actually getting the theory flat-out wrong. The error was always a version of what has been called 'evolutionary panglossism'.

As reported by my friend and mentor the late John Maynard Smith, his own mentor the formidable J. B. S. Haldane satirically coined three erroneous, or at best unreliable, 'theorems'. Aunt Jobiska's Theorem (from Edward Lear) was 'It's a fact the whole world knows . . .' The Bellman's Theorem (from Lewis Carroll) was 'What I tell you three times is true.' And Pangloss's Theorem (from Voltaire) was 'All is for the best in this best of all possible worlds.'

Evolutionary panglossians are vaguely aware that natural selection does a pretty effective job of making living things good at their business of living. Albatrosses seem beautifully designed for flying above the waves, penguins for flying beneath them (I happen to be writing this on a ship in Antarctic waters, marvelling through my binoculars at these prodigies of avian virtuosity). But panglossians forget, as it is so easy to forget, that this 'good at' applies to *individuals*, not to *species*. Good at flying, good at swimming, surviving, reproducing, yes, natural selection will tend to make individual animals good at those things. But there is no reason to expect natural selection to make *species* good at avoiding extinction, good at balancing their sex ratio, good at limiting their population in the interests of the common weal, good at husbanding their food supply and conserving their environment for the benefit of future generations. That would be panglossism. Group survival may emerge as a consequence of improved individual survival, but that is a fortunate by-product. Group survival is not what natural selection is about.

The panglossian error is tempting because we humans are blessed with foresight and can judge which actions are likely to benefit our species, or our town, or our nation, or the whole world, or any specified entity or interest group, in the future. We can foresee that to overfish the seas would be in the long run counterproductive for all fishermen. We can foresee a happier future if we limit our birth rate so that fewer individuals are born, to enjoy richer lives. We can decide that self-restraint now will pay dividends in the future. But natural selection has no foresight.

To be sure, a panglossian version of natural selection theory had been proposed, which, if only it worked, might achieve something like the 'all is for the best' utopia. But unfortunately it doesn't work. At any rate, it was one of my aims in *The Selfish Gene* to persuade my readers that it doesn't work. This was the theory called 'group selection'. This maddeningly seductive error – the Great Group Selection Fallacy or GGSF – ran all through Konrad Lorenz's popular 1964 book *On Aggression*. It also pervaded Robert Ardrey's best-sellers, *The Territorial Imperative* and *The Social Contract* – where I was especially affronted by the mismatch between Ardrey's erroneous message and the high quality of the English in which he expressed it.[1] I aspired to publish a book on the same theme as Ardrey's *Social Contract* (itself a sort of biological rewrite of Rousseau's famous treatise); but mine would be based on rigorous natural selection theory, not the GGSF. My ambition was to undo the damage done by Ardrey and Lorenz – and by many television documentaries of the time, whose promulgation of the error was so ubiquitous that in *The Selfish Gene* I even dubbed it 'the BBC Theorem'.

I was all too familiar with panglossism and the GGSF because I met them weekly in undergraduate essays. And indeed I, when an undergraduate myself, had laced many of my essays with the fallacious view that what really matters in natural selection is the survival of the species (my tutors never noticed). When I eventually came to write *The Selfish Gene*, my dream was that I would change all that. I was daunted by the knowledge that, in order to succeed, my book would need to be as well-written as Ardrey's and sell as prolifically

---

[1] Konrad Lorenz, *On Aggression*, translated by Marjorie Latzke (London, Methuen, 1964); first published in German as *Das sogenannte Böse* – 'the so-called evil' – in 1963. Robert Ardrey, *The Territorial Imperative: A Personal Inquiry into the Animal Origins of Property and Nations* (London: Collins, 1967), and *The Social Contract: A Personal Inquiry into the Evolutionary Sources or Order and Disorder* (London, Collins, 1970).

as Lorenz's. I jokingly spoke of it as 'my best-seller', never really believing that it would become one but giving self-consciously ironic voice to my wilder ambition.

Natural selection is a purely mechanical, automatic process. The world is constantly tending to become full of entities that are good at surviving, denuded of things that are not. Natural selection has no foresight, but brains have, and that is why panglossism is so appealing to us. Brains may agonize about the long-term future and project forward this century's self-indulgence into next century's catastrophe. Natural selection cannot do that. Natural selection can't agonize about anything. Natural selection can only blindly favour short-term gain, because every generation is automatically filled with the offspring of those individuals who did whatever it took, in the short term, to manufacture offspring more effectively than other individuals of their own generation.

And when you look carefully and hard at exactly what is going on as the generations flash by, your gaze is drawn irresistibly to the gene as the level at which natural selection really works. Natural selection automatically favours self-interest among entities that potentially can pass through the generational filter and survive into the distant future. As far as life on this planet is concerned, that means genes. Here's how I put it in *The Selfish Gene*, where I introduced the phrase 'survival machine' to describe the role of (mortal) individual organisms vis-à-vis their (potentially immortal) genes:

> The genes are the immortals . . . [they] have an expectation of life that must be measured not in decades but in thousands and millions of years.
>
> In sexually reproducing species, the individual is too large and too temporary a genetic unit to qualify as a significant unit of natural selection. The group of individuals is an even larger unit. Genetically speaking, individuals and groups are like clouds in the

sky or dust-storms in the desert. They are temporary aggregations or federations. They are not stable through evolutionary time. Populations may last a long while, but they are constantly blending with other populations and so losing their identity. They are also subject to evolutionary change from within. A population is not a discrete enough entity to be a unit of natural selection, not stable and unitary enough to be 'selected' in preference to another population.

An individual body seems discrete enough while it lasts, but alas, how long is that? Each individual is unique. You cannot get evolution by selecting between entities when there is only one copy of each entity! Sexual reproduction is not replication. Just as a population is contaminated by other populations, so an individual's posterity is contaminated by that of his sexual partner. Your children are only half you, your grandchildren only a quarter you. In a few generations the most you can hope for is a large number of descendants, each of whom bears only a tiny portion of you – a few genes – even if a few do bear your surname as well.

Individuals are not stable things, they are fleeting. Chromosomes too are shuffled into oblivion, like hands of cards soon after they are dealt. But the cards themselves survive the shuffling. The cards are the genes. The genes are not destroyed by crossing-over, they merely change partners and march on. Of course they march on. That is their business. They are the replicators and we are their survival machines. When we have served our purpose we are cast aside. But genes are denizens of geological time: genes are forever.

I had already convinced myself of this truth a decade earlier, in almost exactly the same words, when I gave the 1966 undergraduate lectures in Oxford that I have already described. I recalled on page 199 the rhetorical flourishes with which I tried to persuade the undergraduates of the centrality of the immortal gene in the logic of natural selection. Here are my 1966 words, and you can see how similar they are to the equivalent, more rhetorical paragraphs of *The Selfish Gene.*

> Genes are in a sense immortal. They pass through the generations, reshuffling themselves each time they pass from parent to offspring. The body of an animal is but a temporary resting place for the genes; the further survival of the genes depends on the survival of that body at least until it reproduces, and the genes pass into another body . . . the genes build themselves a temporary house, mortal, but efficient for as long as it needs to be . . . To use the terms 'selfish' and 'altruistic' then, our basic expectation on the basis of the orthodox neo-Darwinian theory of evolution is that *Genes will be 'selfish'.*

When I recently found the text of that 1966 lecture (with its encouraging marginal note by Mike Cullen), I was surprised to realize that I had not then read George C. Williams's book, *Adaptation and Natural Selection*, published in the same year:[1]

> With Socrates' death, not only did his phenotype disappear, but also his genotype . . . The loss of Socrates' genotype is not assuaged by any consideration of how prolifically he may have reproduced. Socrates' genes may be with us yet, but not his genotype, because meiosis and recombination destroy genotypes as surely as death.
>
> It is only the meiotically dissociated fragments of the genotype that are transmitted in sexual reproduction, and these fragments are further fragmented by meoisis in the next generation. If there is an ultimately indivisible fragment it is, by definition, 'the gene' that is treated in the abstract discussions of population genetics.

---

[1] George C. Williams, *Adaptation and Natural Selection* (Princeton, NJ, Princeton University Press, 1966).

When I eventually read Williams's great book (some years later, I regret to say), his Socrates passage immediately resonated with me and I prominently acknowledged Williams's importance, as well as Hamilton's, for the theme of *The Selfish Gene* when I came to write it.

Williams and Hamilton were somewhat similar characters: quiet, withdrawn, self-effacing, deep-thinking. Williams had a dignity and mien which – perhaps enhanced by his high forehead and the cut of his beard – reminded many of Abraham Lincoln. Hamilton had more an air of A. A. Milne's Eeyore. But when I wrote *The Selfish Gene* I didn't know either man, just their published work and how central it was for our understanding of evolution.

Because genes are potentially immortal in the form of accurate copies, the difference between successful genes and unsuccessful genes really matters: it has long-term significance. The world becomes filled with genes that are good at being there, good at surviving through many generations. In practice that means good at cooperating with other genes in the business of building bodies that have what it takes to survive long enough to reproduce – for bodies are the temporary vehicles in which genes reside and which pass them on. Throughout *The Selfish Gene* I used that phrase 'survival machine' as my name for an organism. Organisms are the entities in life that actually do things – move, behave, search, hunt, swim, run, fly, feed their young. And the best way to explain everything that an organism does is to assume that it has been programmed, by the genes that ride inside it, to preserve those genes and pass them on before the organism itself dies.

I also used the word 'vehicle' as equivalent to 'survival machine'. It reminds me of an amusing occasion when a Japanese television crew came to interview me about *The Selfish Gene*. They all travelled to Oxford from London packed into a black cab, with tripods and floodlights and, as it seemed, arms and legs sticking out of every window. The director informed me, in halting English (the official interpreter couldn't make me understand him at all and had been sent off in disgrace), that he wanted to film me in the taxi as it drove around Oxford. This puzzled me, and I asked why. 'Hoh!' came the puzzled reply: 'Are you not author of Taxicab Theory of Evolution?' I afterwards guessed that the Japanese translators of my writings must have rendered 'vehicle' as 'taxicab'.

The interview itself was quite amusing. I rode in the taxi alone except for the cameraman and sound man. In the absence of the official interpreter, there was no interviewer, and I was bidden simply to talk *ad lib* about *The Selfish Gene* while we did a scenic tour of Oxford. The taxi driver doubtless had the streets of London intricately mapped in his enlarged hippocampus, but he didn't know Oxford. It therefore fell to me to guide him, and my otherwise measured discourse on selfish genes was punctuated by frantic shouts of 'Turn left here!' or 'Turn right at the traffic lights and then get in the right-hand lane!' I hope they managed to find the unfortunate interpreter before returning to London.

In *The Selfish Gene*, I criticized the panglossian idea that animals have foresight and work out what would be good for the long-term future of their species or group. What is wrong with this is not the idea that animals 'work out what would be good'. There is in any case no suggestion that the 'working out' is conscious. No: what is wrong is the idea of the species or group as the entity whose benefit is maximized. Biologists often legitimately use the language of 'working out what would be good' as a shorthand route to sound Darwinian reasoning. The trick is to identify the correct level in the hierarchy of life at which the shorthand metaphor of conscious reasoning is applied. It is quite OK to put yourself in the position of an individual animal and ask: 'What would I do if I were trying to achieve the goal of propagating my genes?'

The Selfish Gene is filled with imagined soliloquies in which a hypothetical animal 'reasons' to itself: 'Should I do X or Y?' The meaning of 'should' is: 'Would X or Y be better for my genes?' This is legitimate, but only because it can be translated into the question: 'Would a gene for making individuals do X (in this situation) become more frequent in the gene pool?' The subjective soliloquy is justified by the fact that it can be translated into the language of gene survival.

One might be tempted to interpret 'Should I do X or Y?' as meaning 'Would X or Y be more likely to prolong my own life?' But if long life is bought at the expense of not reproducing – that is to say, if we pit individual longevity against gene survival – natural selection will not favour it. Reproduction can be a dangerous business. Male pheasants that are gorgeously coloured to attract females also attract predators. A drab, inconspicuous male would probably live longer than a brightly coloured, attractive male. But he'd be more likely to die unmated, and the genes for safety-first drabness are less likely to be passed on. Gene survival is what really matters in natural selection.

The following is a legitimate shorthand, put into the mouth of a male pheasant: 'If I grow drab feathers I shall probably live a long time, but I won't get a mate. If I grow bright feathers I'll probably die young, but I'll pass on lots of genes before I die, including the genes for making bright feathers. Therefore I should take the "decision" to grow bright feathers.' Needless to say, 'decision' doesn't mean what a human would ordinarily mean by it. Conscious thought is not involved. Organism-level shorthand can be confusing, but it works so long as you remember always to keep open the pathway to a translation back into gene language. No pheasant actually takes a 'decision' to grow bright feathers or drab. Instead, genes for growing bright feathers or drab have different probabilities of surviving through the generations.

It really can be helpful, when trying to understand from a

Darwinian point of view what animals do, to see them as robot machines, 'thinking' about what steps to take in order to pass on their genes to future generations. Such steps may involve behaving in certain ways, or growing organs of a particular shape or character. It can also be helpful to think metaphorically of *genes* as 'thinking' about what steps to take in order to pass themselves on to future generations. Such steps will usually involve manipulating individual organisms via the processes of embryonic development.

But it is never even *metaphorically* legitimate to treat animals as thinking about what steps to take in order to preserve their species, or their group. Differential group or species survival is not what happens in natural selection. What happens is differential gene survival. Therefore, legitimate shorthands are of the form 'If I were a gene, what would I do to preserve myself?' Or – and ideally this should be exactly equivalent – 'If I were an organism, what would I do to preserve my genes?' But 'If I were an organism, what would I do to preserve my species?' is an illegitimate shorthand. So is – this time for a different reason – 'If I were a species, what would I do to preserve myself?' The latter metaphor is illegitimate because a species, unlike an individual organism, is not the kind of entity that even metaphorically behaves as an agent, doing things, acting upon decisions. Species don't have brains and muscles, they are just collections of individual organisms that do. Species and groups are not 'vehicles'. Individual organisms are.

I should point out that neither in my lectures of the 1960s nor in *The Selfish Gene* did I see as very novel the idea of the gene as the fundamental unit of natural selection. I thought of it – and clearly said so – as implicit in the orthodox neo-Darwinian theory of evolution: that is to say, the theory first clearly formalized in the 1930s by Fisher, Haldane, Wright and the other founding fathers of the so-called Modern Synthesis such as Ernst Mayr, Theodosius Dobzhansky, George Gaylord Simpson and Julian Huxley. It was only after *The Selfish Gene* was published that both critics and admirers came to see the idea as revolutionary. That was not how I thought of it at the time.

Having said that, however, I should add that not all the founding fathers of the Modern Synthesis were clear about this important implication of the theory that they collectively put together. To the end of his centenarian life, the authoritative German-American taxonomist Ernst Mayr expressed hostility to the idea of gene selectionism, in terms that suggested to me that he misunderstood it. And Julian Huxley, the founding father who actually coined the phrase 'Modern Synthesis', was an out-and-out group selectionist, without clearly realizing it. The first time I met the great Peter Medawar, he startled my student self with a deliciously sacrilegious remark, delivered with his characteristically patrician, yet impish style. 'The trouble with Julian is that he really doesn't *understand* evolution.' Fancy saying that – of a Huxley! I could hardly believe my ears and, as you see, I have never forgotten it. I later heard another Nobel Prize-winner, the French molecular biologist Jacques Monod, say something a bit similar, though not about Huxley: 'The trouble with natural selection is that everybody *thinks* he understands it.'

I mentioned that I began *The Selfish Gene* when power cuts interrupted my cricket research. I had completed only the first chapter of the book when I happened to meet an editor from the publishers Allen & Unwin. He was paying a routine visit to the Department of Zoology in search of possible books, and I told him about my embryonic project. He sat down and read that first chapter on the spot, liked it and encouraged me to continue. But then – unfortunately from one narrow point of view, fortunately from others – the industrial unrest came to an end and the lights came back on. I shoved my chapter in a drawer and forgot about it as I resumed my research on crickets.

From time to time during the next two years I contemplated

returning to the book. The impetus was especially strong when I read and lectured about new publications that were beginning to appear in the early 1970s and proved beautifully compatible with the thesis of my gestating book. Most notable among these were papers by the young American biologist Robert Trivers, and others by the veteran British professor John Maynard Smith. Both these authors made use of the intuitive shortcut I mentioned (the philosopher Daniel Dennett would now call it an intuition pump):[1] the shortcut of imagining that an individual organism behaves 'as if' consciously calculating the best policy for preserving and propagating its genes.

Trivers treated a parent animal *as if* it were a rational agent calculating what economists call the 'opportunity cost' of an action. A parent has to pay the costs of rearing each offspring. Among these costs might be food, including time and effort spent gathering it, time spent protecting the child from predators, and the risks incurred by the parent in doing so. Trivers wrapped them all up in one metric which he called Parental Investment or PI. Trivers' key insight was that that PI must be an *opportunity cost*: the investment in any one child is measured as *lost opportunities to invest in other children*. Trivers used the notion to develop a penetrating theory of 'parent–offspring conflict'. The decision on the best time to wean a child, for example, is subject to a 'disagreement' between the child and its mother, both behaving as rational economists whose 'utility function' is the long-term survival of their own genes. The mother 'wants' to terminate suckling earlier than the child does, because she places greater 'value' than he does on her future offspring, who will benefit from early weaning of the present child. The present child also 'values' his future siblings, but only half as highly as his mother does because of the way Hamilton's Rule pans out. Therefore there

---

[1] Daniel C. Dennett, *Intuition Pumps and Other Tools for Thinking* (New York, Norton, 2013).

is a period of 'weaning conflict', an uneasy phase of transition between the early time when both parties 'agree' that suckling should continue and the later time when both parties 'agree' that it should end. During this phase, when the mother 'wants' weaning but the child doesn't, observers of animal behaviour should see the symptoms of a subtle battle between mother and child. In passing I should add that, long after *The Selfish Gene* was published, the Australian biologist David Haig cleverly showed how many of the ailments of pregnancy can be explained in terms of the same Triversian conflict going on inside the womb – not about weaning in this case, obviously, but about other aspects of the allocation of necessarily scarce resources.

Parent–offspring conflict was obviously a subject tailor-made for my book, and Trivers' brilliant paper on the subject was one of the spurs that encouraged me to take my first chapter out of the drawer where it had languished since the end of the power strike. It became the inspiration for chapter 7 of *The Selfish Gene*, 'Battle of the generations'. Chapter 8, 'Battle of the sexes', also made use of Trivers' ideas, this time showing how males and females might compute their opportunity costs differently. When might a male, for example, desert his mate, leaving her 'holding the baby' and 'in a cruel bind' while he seeks a new one? Trivers also influenced chapter 10, 'You scratch my back, I'll ride on yours'. His paper in this case was an earlier one, on reciprocal altruism, which showed that kin selection is not the only evolutionary pressure towards altruism. Reciprocation – the repaying of favours – can also be very important, and it works across species, not just within them as kin selection does. So Trivers' name was added to those of Hamilton and Williams among the four authors who had greatest influence on *The Selfish Gene*. I also asked him to write the foreword – which he graciously did, although at that point we had never met.

The fourth was John Maynard Smith, who later became a

beloved mentor. As a boy I had met the book that he would refer to as 'my little Penguin', and I was much taken by the smiling author photograph: shock of nutty professor hair askew like the pipe in his mouth, thick round glasses in need of a clean – the sort of man I was immediately drawn to. I also liked the biographical note, which explained that he had been an engineer designing aircraft, but gave it up and went back to university to study biology because he noticed that 'aircraft were noisy and old fashioned'. Many years later, a new edition of that book, *The Theory of Evolution*, was published by Cambridge University Press and I was honoured to be invited to write the foreword.[1] I included the following tribute to this genial hero:

> Readers of 'campus novels' know that a conference is where you can catch academics at their worst. The conference bar, in particular, is the academy in microcosm. Professors huddle together in exclusive, conspiratorial corners, talking not about science or scholarship but about 'tenure-track hiring' (their word for jobs) and 'funding' (their word for money). If they do talk shop, too often it will be to make an impression rather than to enlighten. John Maynard Smith is a splendid, triumphant, lovable exception. He values creative ideas above money, plain language above jargon. He is always the centre of a lively, laughing crowd of students and young research workers of both sexes. Never mind the lectures or the 'workshops'; be blowed to the motor coach excursions to local beauty spots; forget your fancy visual aids and radio microphones; the only thing that really matters at a conference is that John Maynard Smith must be in residence and there must be a spacious, convivial bar. If he can't manage the dates you have in mind, you must just reschedule the conference. He doesn't have to give a formal talk (although he is a riveting speaker) and he doesn't have to chair a formal session (although

---

[1] John Maynard Smith, *The Theory of Evolution* (Cambridge, Cambridge University Press, 1993; first published London, Penguin, 1958).

he is a wise, sympathetic and witty chairman). He has only to turn up and your conference will succeed. He will charm and amuse the young research workers, listen to their stories, inspire them, rekindle enthusiasms that might be flagging, and send them back to their laboratories or their muddy fields, enlivened and invigorated, eager to try out the new ideas he has generously shared with them.

My relationship with John didn't get off to an outstandingly good start, however. I first met him in 1966 when, as Dean of Biological Sciences, he interviewed me for a job at the University of Sussex. I was already pretty much committed to going to Berkeley. However, there was a job going at Sussex and Richard Andrew, their resident expert on our shared subject of animal behaviour, pressed me with flattering urgency to apply. I told Richard about my near-commitment to go to Berkeley, but he said there was no harm in doing the Sussex interview anyway, so I thought: What the hell, why not? I'm afraid my 'what the hell' attitude didn't endear me to Maynard Smith in the interview. I said I wouldn't lecture about animal taxonomy. He said it was part of the job. I rather arrogantly said: Well, I've got a job offer at Berkeley and I'm not quite sure why I am doing this interview anyway. He was nice about it when he and Dr Andrew took me to lunch, but, as I said, it was not a good start to what was later to prove a delightful friendship.

In the early 1970s, Maynard Smith began the long series of papers in which, together with colleagues such as Geoffrey Parker and the late George Price, he deployed a version of the mathematical theory of games to solve a number of problems in evolution. These ideas were immensely congenial to the idea of the selfish gene, and Maynard Smith's papers constituted the other major stimulus that led me to dust off my old chapter 1 and write the whole book.

Maynard Smith's particular contribution was the notion of the evolutionarily stable strategy or ESS. 'Strategy' in this sense can be construed as 'preprogrammed rule'. Maynard Smith set up mathematical models in which preprogrammed rules with names like (for the particular case of animal combat) Hawk, Dove, Retaliator, Bully, are let loose in an imagined (or simulated) world to interact with each other. Once again, it is important to understand that the animals implementing the rules are not assumed to be consciously aware of what they are doing, or why. Each preprogrammed rule has a *frequency* in the population (like genes in a gene pool, although the link with DNA doesn't have to be made explicit in the models). The frequencies change in accordance with 'payoffs'. In the social and economic sciences where game theory originated, payoffs can be thought of as equivalent to money. In evolutionary game theory, payoffs have the special meaning of reproductive success: high payoffs for a strategy lead to increased representation in the population.

The key point is that a successful strategy is not necessarily one that wins its particular contests against other strategies. A successful strategy is one that numerically dominates the population. And since a numerically dominant strategy is by definition likely to encounter copies of *itself*, it will stay numerically dominant only if it flourishes in the presence of copies of itself. This is the meaning of 'evolutionarily stable' in Maynard Smith's 'ESS'. We expect to see ESSs in nature, because if a strategy is evolutionarily *un*stable, it will tend to disappear from the population as rival strategies outbreed it.

I won't expound evolutionary game theory any further here because I did that in *The Selfish Gene*, and the same applies to Trivers' ideas on parental investment. Here it is sufficient to say that the publications of Trivers and Maynard Smith in the early 1970s rekindled my interest in the ideas of Hamilton that had inspired me in the 1960s, and moved me to return to the book whose first chapter had slumbered in a drawer since the end of the power strikes. Maynard Smith's game-theoretic ideas dominated the chapter on aggression, and inspired my treatment of many topics in later chapters.

So, finally, in 1975, having finished my 'hierarchical organization' paper, I took the sabbatical leave to which I was entitled, stayed at home every morning, and devoted myself to my typewriter and *The Selfish Gene*. Indeed, so devoted was I to the task that I didn't attend the crucial meeting at which New College was electing a new Warden. A colleague slipped out of the meeting and telephoned urgently to tell me the vote was extremely close and to beg me to come quickly. I now think that, although my sabbatical leave entitled me to do so, my absenting myself from such a crucial vote was an act of self-indulgent irresponsibility. The meeting would have taken only a few hours of my time, and the repercussions of my lost vote potentially might have been felt for many years. Fortunately, the man for whom I would have voted got in anyway (and became an excellent Warden), so I don't have to endure a burden of guilt for changing the course of college history. Actually, his rival would have been very good too, and college meetings would certainly have been amusing as he was justly reputed to be the wittiest man in Oxford.

I wrote *The Selfish Gene* in a frenzy of creative energy. I'd completed three or four chapters when I spoke to my friend Desmond Morris about publication. A legendarily successful author himself, Desmond arranged a meeting with Tom Maschler, doyen of London publishers. I met Mr Maschler in his high-ceilinged, book-lined room at Jonathan Cape in London. He'd read my chapters and liked them, but urged me to change the title. 'Selfish', he explained to me, is a 'down word'. Why not *The Immortal Gene*? With hindsight, he was very probably right. I can't now remember why I didn't follow his advice. I think I should have done.

I didn't, in any case, pursue him as a publisher because matters were rather forcefully taken out of my hands. At lunch one day in New College Roger (now Sir Roger) Elliott, Oxford's Professor of Theoretical Physics, said he had heard I was writing a book, and asked me about it. I told him a little of what I was trying to do, and he seemed interested. As it happened, he was a member of the Board of Delegates of Oxford University Press, and he passed the word to Michael Rodgers, the appropriate editor at that ancient publishing house. Michael wrote and asked to see my chapters. I sent them to him.

And then the whirlwind hit – beginning with his characteristically loud voice, over the telephone: 'I've read your chapters. Haven't been able to sleep since. *I MUST HAVE THAT BOOK!*' Well, some people might resist that kind of persuasion, but not me. Michael was clearly my kind of publisher. I signed the contract and set to work with redoubled urgency to complete the book.

I now find it quite hard to comprehend how we all used to tolerate the burden of writing in the age before computer word processors. Pretty much every sentence I write is revised, fiddled with, re-ordered, crossed out and reworked. I reread my work obsessively, subjecting the text to a kind of Darwinian sieving which, I hope and believe, improves it with every pass. Even as I type a sentence for the first time, at least half the words are deleted and changed before the sentence ends. I have always worked like this. But while a computer is naturally congenial to this way of working, and the text itself remains clean with every revision, on a typewriter the result was a mess. Scissors and sticky tape were tools of the trade as important as the typewriter itself. The growing typescript of *The Selfish Gene* was covered with xxxxxxx deletions, handwritten insertions, words ringed and moved with arrows to other places, strips of paper inelegantly taped to the margin or the bottom of the page. One would think it a necessary part of composition that one should be able to read one's text fluently. This would seem to be impossible when working on paper. Yet, mysteriously, writing style does not seem to have shown any general improvement since the introduction of computer word processors. Why not?

The Selfish Gene went through two fair copies typed up by Pat Searle, the motherly secretary of the Animal Behaviour Research Group. Each one went to Michael Rodgers and came back with his helpful, handwritten annotations. In particular, he excised some purple passages that my romantically youthful enthusiasm had pushed way over the top. In Peter Medawar's metaphor of the writer as organist, 'a scientist's fingers, unlike a historian's, must never stray toward the diapason'. The end of chapter 2 of *The Selfish Gene* is about as purple as science prose should get, and I blush to recall (and am glad I haven't preserved) the paragraph that followed it. Here's the paragraph of paler purple that survived Michael's moderating pen. It's the end of the chapter on the origin of life and the spontaneous arising in the primeval soup of 'replicators', which later moved into the world of 'vehicles' – living organisms.

Was there to be any end to the gradual improvement in the techniques and artifices used by the replicators to ensure their own continuation in the world? There would be plenty of time for improvement. What weird engines of self-preservation would the millennia bring forth? Four thousand million years on, what was to be the fate of the ancient replicators? They did not die out, for they are past masters of the survival arts. But do not look for them floating loose in the sea; they gave up that cavalier freedom long ago. Now they swarm in huge colonies, safe inside gigantic lumbering robots, sealed off from the outside world, communicating with it by tortuous indirect routes, manipulating it by remote control. They are in you and in me; they created us, body and mind; and their preservation is the ultimate rationale for our existence. They have come a long way, those replicators. Now they go by the name of genes, and we are their survival machines.

That paragraph encapsulates the central metaphor of the book, and also its science-fictiony feel. Indeed, I began my preface with the words:

> This book should be read almost as though it were science fiction. It is designed to appeal to the imagination. But it is not science fiction: it is science. Cliché or not, 'stranger than fiction' expresses exactly how I feel about the truth. We are survival machines – robot vehicles blindly programmed to preserve the selfish molecules known as genes. This is a truth which still fills me with astonishment. Though I have known about it for years, I never seem to get fully used to it. One of my hopes is that I may have some success in astonishing others.

And the opening lines of chapter 1 continued the science fiction mood:

> Intelligent life on a planet comes of age when it first works out the reason for its own existence. If superior creatures from space ever visit earth, the first question they will ask, in order to assess the level of our civilization, is: 'Have they discovered evolution yet?' Living organisms had existed on earth, without ever knowing why, for over three thousand million years before the truth finally dawned on one of them. His name was Charles Darwin.

Niko Tinbergen hated that opening, when the book was published and he read it. He didn't like anything that suggested that humanity is an intelligent species, and he felt deeply wounded by the terrible effects we have had on the world. But that really wasn't the kind of point I was making.

I should say something about the last chapter: 'Memes: the new replicators'. Given that the rest of the book thrust the gene to centre stage as the starring replicator in the evolution of life, it was important to dispel the impression that the replicator has to be DNA. In keeping with the science fiction mood of the opening, I pointed out that on other planets the evolution of life could be fostered by a completely different system of self-replication – but that, whatever it was, it would have to have certain qualities, such as high fidelity of copying.

Casting around for an example, I could have used computer viruses if they had been invented in 1975. Instead, I lit upon human culture as a new 'primeval soup':

> But do we have to go to distant worlds to find other kinds of replicator and other, consequent kinds of evolution? I think that a new kind of replicator has recently emerged on this very planet. It is staring us in the face. It is still in its infancy, still drifting clumsily about in its primeval soup, but already it is achieving evolutionary change at a rate which leaves the old gene panting far behind.
>
> The new soup is the soup of human culture. We need a name for the new replicator, a noun which conveys the idea of a unit of cultural transmission, or a unit of *imitation*. 'Mimeme' comes from a suitable Greek root, but I want a monosyllable that sounds a bit like 'gene'. I hope my classicist friends will forgive me if I abbreviate mimeme to *meme*. If it is any consolation, it could alternatively be thought of as being related to 'memory', or to the French word *même*. It should be pronounced to rhyme with 'cream'.
>
> Examples of memes are tunes, ideas, catch-phrases, clothes fashions, ways of making pots or of building arches. Just as genes propagate themselves in the gene pool by leaping from body to body via sperms or eggs, so memes propagate themselves in the meme pool by leaping from brain to brain, via a process which, in the broad sense, can be called *imitation*.

I went on to discuss various ways in which the idea of memes might be applied, for example to the spread and inheritance of

religion. My primary intention, however, was not to make a contribution to the theory of human culture, but to downplay the gene as
the only conceivable replicator that might lie at the root of a
Darwinian process. I was trying to push 'Universal Darwinism' (the
title of a later paper, based on my lecture to the 1982 conference
commemorating Darwin's death). Nevertheless I am delighted that
the philosopher Daniel Dennett, the psychologist Susan Blackmore
and others have run, so productively, with the meme ball. More than
thirty books have now been published with the word 'meme' in their
title, and the word has made it into the *Oxford English Dictionary*
(whose criterion is that it must be used, without attribution or
definition, in a significant number of published places).

Publication of one's first book is a heady time for a young
author. I made frequent trips to the stately neo-classical OUP building in Walton Street, and sometimes the London office in Ely House,
to meet the various people involved in the complex business of production, design, marketing and so on. When it came to jacket
design, the science fiction mood of the book led me again to the
elegantly porticoed North Oxford door of Desmond Morris. As well
as being a biologist, television personality, anthropological collector,
(implausible) raconteur[1] and best-selling author, Desmond is an
accomplished surrealist painter. His paintings have an unmistakably
biological feel. He has created a dreamscape in which other-worldly
creatures live and move and have their evolution – for they do
evolve, from canvas to canvas: just what was needed for *The Selfish
Gene*. He was delighted by the idea of providing a jacket design, and
Michael Rodgers and I went to look at the paintings on his walls

---

[1] I suspect him of being the original source of a widely circulated anecdote about the film star
Diana Dors. She and he came from the same Wiltshire town and were childhood friends. Her
real surname was not Dors but Fluck. She was invited back to open some fête or other, and the
vicar, thinking to introduce her by the name the locals would have known, genially asked them
to welcome the lovely 'Diana . . . Clunt'.

and in his studio. *The Expectant Valley* stood out, not just for its bold colours and air of brooding fecundity, but also more mundanely in that it provided a convenient space to accommodate the title. We chose it with pleasure and I believe it enhanced the sales of the book.

As it happened, Desmond had an exhibition around this time in a small gallery in Walton Street near the OUP building, and *The Expectant Valley* was one of the paintings on sale. Its price, £750, happened to be exactly equal to the advance the publishers had paid me for my book. The coincidence was too much to resist and, after repeated visits to the gallery during which I became fond of many of the paintings, I bought *The Expectant Valley*. I think Desmond was a bit embarrassed, and he kindly threw in another, slightly similar painting, *The Titillator*. The two go rather well together.

The Selfish Gene was published in the autumn of 1976; I was thirty-five. It was reviewed widely, surprisingly so for a first work by an unknown author, and I still don't really know why it received the attention it did. There was no launch party and no obvious fanfare organized by the publishers. Some months after publication it came to the notice of Peter Jones, one of the producers on the BBC's 'flagship' science series *Horizon*. Peter asked me if I would like to present a documentary on the subject, but I was much too shy at that time to dare appear on television, and I recommended John Maynard Smith instead. He did a good job – he had a wonderfully warm and engaging manner – and the documentary, which had the same name, *The Selfish Gene*, must have given a good boost to the book's sales, at least in Britain. But the broadcast came too late to account for the wide review coverage the book received.

I don't do it any more, but for that first book I kept a scrap-book of reviews, and I have just been glancing at them again. There were more than 100, and a rereading doesn't generally bear out the common perception of the book as controversial. Almost all the reviews were favourable. Among early reviewers were the psychiatrist Anthony Storr, the anthropologists Lionel Tiger and Francis Huxley (son of Julian), the naturalist Bruce Campbell and the philosopher Bernard Williams, whom I came to know much later as one of those entertaining conversationalists whose wit had the capacity to 'raise the game' of any companion. There were hostile reviews from two biologists identified with the political left, Steven Rose and Richard Lewontin, and – more subtly barbed – from Cyril Darlington on the opposite side of the political spectrum. But these were rare. Most of the reviewers got the message, expounded it fairly and were very nice about the book. Especially warming for me were the highly favourable reviews by Peter Medawar and W. D. Hamilton. Hamilton even hit the particular nail I had originally targeted in my quest to answer Lorenz, Ardrey and the panglossians of the 1960s and the 'BBC Theorem':

> This book should be read, can be read, by almost everybody. It describes with great skill a new face of the theory of evolution. With much of the light, unencumbered style that has lately sold new and sometimes erroneous biology to the public, it is, in my opinion, a more serious achievement. It succeeds in the seemingly impossible task of using simple, untechnical English to present some rather recondite and quasi-mathematical themes of recent evolutionary thought. Seen through this book in their broad perspective at last, these will surprise and refresh even many research biologists who might have supposed themselves already in the know. At least, so they surprised this reviewer. Yet, to repeat, the book remains easily readable by anyone with the least grounding in science.

There was nobody in the world whom I would rather have surprised in such a way than 'this reviewer'. I was also touched by the way Bill Hamilton ended his beautifully written review with poems, one by Wordsworth, and the other by Housman, whose Shropshire

Lad I often found myself identifying with Bill's complex personality:

> From far, from eve and morning
> And yon twelve-winded sky,
> The stuff of life to knit me
> Blew hither: here am I

> . . .

> Speak now, and I will answer;
> How shall I help you, say;
> Ere to the wind's twelve quarters
> I take my endless way.

Not a bad epitaph for an evolutionary scientist, and Bill Hamilton was probably the greatest evolutionary scientist of the second half of the twentieth century. While this volume of auto-biography was in its final stages, I found a treasure among a bundle of old papers, with Bill's handwriting at the top: it was a copy of the last page of his lecture notes, containing a rewriting of another Housman poem, 'The Immortal Part', to incorporate the idea of the 'immortal' gene. I have no memory of the lecture he is referring to, or when he gave it, and the paper is undated. I have reproduced it in the web appendix.

Long after *The Selfish Gene* was published, Bill became my close colleague at Oxford and I saw him almost daily at lunch in New College. I am humbly proud of the role played by my book in bringing his brilliant ideas to a wider audience. But I like to hope, too, that there are other ways in which the book changed the way my professional colleagues think about their subject. I like to think it is no accident that, if you visit a biological field station in the Serengeti, or Antarctica, or the Amazon or the Kalahari, and listen to the active researchers talking shop over their beers in the evenings, what you hear will be laced with talk of genes. It's not that they're talking about the molecular antics of DNA – although that is interesting too – but the underlying assumption of these conversations is that the behaviour of the animals and plants under study is aimed at preserving genes and propagating them through succeeding generations.

LOOKING BACK
DOWN THE PATH

PUBLICATION of *The Selfish Gene* marks the end of the first half of my life and a convenient point at which to pause and look back. I'm often asked whether my African childhood led me to become a biologist. I'd like to answer yes, but I'm not confident. How can we know whether the course of a life would have been changed by some particular alteration in its early history? I had a trained botanist for a father and a mother who knew the name of every wildflower you could normally expect to see – and both of them were always eager to satisfy a child's curiosity about the real world. Was that important to my life? Yes, it surely was.

My family moved to England when I was eight. What if they hadn't? At the eleventh hour I was sent to Oundle rather than Marlborough. Did that arbitrary change seal my future? Both were boys-only schools. A psychologist might suggest that I'd have turned out a socially better-adjusted person if I'd been sent to mixed schools. I scraped into Oxford. What if I'd failed, as probably I nearly did? What if I had never had tutorials with Niko Tinbergen, and therefore followed my earlier plan to do biochemical research for my DPhil rather than animal behaviour? Surely my whole life would then have been different? Probably I would never have written any books.

But perhaps life has a tendency to converge on a pathway, something like a magnetic pull that draws it back despite temporary deviations. As a biochemist, might I have eventually returned to the path that led to *The Selfish Gene*, even if I had then given it a more molecular slant? Perhaps the pull of the pathway would have led me to write (again biochemically slanted) versions of every one of my dozen books. I doubt it, but this whole 'returning to the path' idea is not uninteresting and I shall . . . er . . . return to it.

The hypotheticals that I posed are relatively large. Take something utterly trivial yet, I shall argue, momentous. I've already speculated that we mammals owe our existence to a particular sneeze by a particular dinosaur. What if Alois Schicklgruber had happened to sneeze at a particular moment – rather than some other particular moment – during any year before mid-1888 when his son Adolf Hitler was conceived? Obviously I have not the faintest idea of the exact sequence of events involved, and there are surely no historical records of Herr Schicklgruber's sternutations, but I am confident that a change as trivial as a sneeze in, say, 1858 would have been more than enough to alter the course of history. The evil-omened sperm that engendered Adolf Hitler was one of countless billions produced during his father's life, and the same goes for his two grandfathers, and four great-grandfathers, and so on back. It is not only plausible but I think certain that a sneeze many years before Hitler's conception would have had knock-on effects sufficient to derail the trivial circumstance that one particular sperm met one particular egg, thereby changing the entire course of the twentieth century including my existence. Of course, I'm not denying that something like the Second World War might well have happened even without Hitler; nor am I saying that Hitler's evil madness was inevitably ordained by his genes. With a different upbringing Hitler might have turned out good, or at least uninfluential. But certainly his very existence, and the war as it turned out, depended upon the fortunate – well, unfortunate – happenstance of a particular sperm's luck.

> A million million spermatozoa,
> All of them alive:
> Out of their cataclysm but one poor Noah
> Dare hope to survive.
>
> And among that billion minus one
> Might have chanced to be
> Shakespeare, another Newton, a new Donne–
> But the One was Me.
>
> Shame to have ousted your betters thus,
> Taking ark while the others remained outside!
> Better for all of us, froward Homunculus,
> If you'd quietly died!
>
> Aldous Huxley

If his father had sneezed at a particular hypothetical moment, Adolf Hitler would not have been born. Nor would I, for I owe my improbable conception to the Second World War – as well as to much less momentous things that happened. And of course all of us can take the argument back through countless previous generations, as I did with my hypothetical dinosaur and the destiny of the mammals.

Taking on board the contingent frailty of the event chain that led to our existence, we can still go on to ask – as I did a moment ago – whether the course of a named individual's life is sucked back, magnetically, into predictable pathways, despite the Brownian buffetings of sneezes and other trivial, or not so trivial, happenings. What if my mother's joking speculation were really true, if the Eskotene Nursing Home really had muddled me up with Cuthbert's son and I had been brought up as a changeling in a missionary household? Would I now be an ordained missionary myself? I think geneticists know enough to say no, probably no.

If my family had stayed on in Africa and I had persisted at Eagle rather than moving to Chafyn Grove, and then been sent to Marlborough rather than Oundle, would I still have got into Oxford and met Niko Tinbergen? It is not unlikely, for my father would have been hell-bent on my following him and half a dozen earlier Dawkinses into Balliol. Despite taking other forks in the road, pathways can converge again. The likelihood that they will do so depends on genuinely investigable matters such as the relative contribution of genes and education to adult abilities and proclivities.

We can leave rarefied speculations about hypothetical sneezes and converging pathways, and return to familiar territory. As a man looks back on his life so far, how much of what he has achieved, or failed to achieve, could have been predicted from his childhood? How much can be attributed to measurable qualities? To the interests and pastimes of his parents? To his genes? To his happening to meet a particularly influential teacher, or happening to go to summer camp? Can he list his abilities and shortcomings, his pluses and minuses, and use them to understand his successes and failures? This is the familiar territory I meant, and it is that trodden, for example, by Darwin at the end of his autobiography.

Charles Darwin is my greatest scientific hero. Philosophers are fond of saying that all philosophy is a series of footnotes to Plato. I sincerely hope that is not the case, because it doesn't say much for philosophy. A far better case could be made that all of modern biology is a series of footnotes to Darwin. And that would be a genuine compliment to the science of biology. Every biologist treads in Darwin's footsteps and, in all humility, none of us could do better than to follow his example. In the closing pages of his autobiography he essayed a retrospective itemization of personal faculties lacked or possessed. Again in humility I shall do the same, taking his method of self-assessment as a model to be followed.

> ... I have no great quickness of apprehension or wit which is so remarkable in some clever men, for instance Huxley.

Here, at least, I can claim mental kinship with Darwin, although in his case the modesty was exaggerated.

> My power to follow a long and purely abstract train of thought is very limited; I should, moreover, never have succeeded with metaphysics or mathematics.

Again, same for me, despite the ludicrously ill-founded reputation for mathematical ability that I briefly enjoyed – or endured – in Bevington Road days. John Maynard Smith, as a mathematical biologist himself, engagingly expressed wonderment at how it is possible to 'think in prose'. He said it in the *London Review of Books* in 1982, at the end of a joint review of *The Selfish Gene* and its sequel (aimed at professional biologists), *The Extended Phenotype*:

> I have left till last what is to me the strangest feature of both books, because I suspect it will not seem strange to many others. It is that neither book contains a single line of mathematics, and yet I have no difficulty in following them, and as far as I can detect they contain no logical errors. Further, Dawkins has not first worked out his ideas mathematically and then converted them into prose: he apparently thinks in prose, although it may be significant that, while writing *The Selfish Gene*, he was recovering from a severe addiction to computer programming, an activity which obliges one to think clearly and to say exactly what one means. It is unfortunate that most people who write about the relation between genetics and evolution without the intellectual

prop of mathematics are either incomprehensible or wrong, and not infrequently both. Dawkins is a happy exception to this rule.

Back to Darwin's autobiographical soliloquy:

So poor in one sense is my memory, that I have never been able to remember for more than a few days a single date or a line of poetry.

That might well have been really true of Darwin, and it doesn't seem to have held him back. My ability to remember poetry word for word hasn't helped my science much, although it has enriched my life and I would not ever wish to lose it. Perhaps, too, a feeling for poetic cadence has some influence on writing style.

My habits are methodical, and this has been of not a little use for my particular line of work. Lastly, I have had ample leisure from not having to earn my own bread. Even ill-health, though it has annihilated several years of my life, has saved me from the distractions of society or amusement.

My habits are anything but methodical, and that – not ill health in my case – has surely annihilated what might have added up to years of more productive life. The same accusation could be levelled at the distractions of society or amusement (and playing with computers in my case), but life is for living as well as producing. I have had to earn my own bread. But – while happy to ignore the attacks I have (yes, really) received for being white, male and adequately educated – I cannot deny a measure of unearned privilege when I compare my childhood, boyhood and youth to others less fortunate. I do not apologize for that privilege any more than a man should apologize for his genes or his face, but I am very conscious of it. And I am grateful to my parents for giving me what will strike some as a favoured childhood. Others might consider it less than a blessing to have been sent away to the spartan regime of boarding school aged seven, but even there I have reason to be grateful to my parents, for whom this style of education was a great expense, necessitating sacrifices from them.

Darwin had earlier let his modesty guard drop a little when he considered his – by any standards formidable – powers of reasoning:

> Some of my critics have said, 'Oh, he is a good observer, but has no power of reasoning.' I do not think this can be true, for the *Origin of Species* is one long argument from the beginning to the end, and it has convinced not a few able men. No one could have written it without having some power of reasoning.

Mr Darwin (never Sir Charles, and what an amazing indictment of our honours system that is), that last sentence should win a prize for world-class understatement. Mr Darwin, you are one of the great reasoners and one of the great persuaders of all time.

I am not a good observer. I'm not proud of it and I try eagerly, but I am not the naturalist my father and his father would have wished. I lack patience and have no great knowledge of any particular animal or – despite one privilege of my upbringing – plant group. I know the songs of only half a dozen common British songbirds, and can recognize only about the same number of constellations in our night sky or families of our wildflowers. I am much better at the phyla, classes and orders of the animal kingdom – and so I should be, having studied zoology at Oxford: for no other university placed such an emphasis on that classical approach to the subject.

The evidence suggests that I am a reasonably effective persuader. Needless to say, the subjects about which I persuade are small beer compared to Darwin's – except in the sense that, amazingly, the job of persuading people of Darwin's own truth is still not over, and I am one of the labourers in Darwin's vineyard today. But that story belongs in the second half of my life, during which the majority of my books were written: it belongs in the companion volume that should follow in two years' time – if I am not carried off by the unpredictable equivalent of a sneeze.

ACKNOWLEDGEMENTS

For advice, help and support of various kinds, I would like to thank Lalla Ward Dawkins, Jean Dawkins, Sarah and Michael Kettlewell, Marian Stamp Dawkins, John Smythies, Sally Gaminara, Hilary Redmon, Sheila Lee, Gillian Somerscales, Nicholas Jones, John Brockman, David Glynn, Ross and Christine Hildebrand, Bill Newton Dunn, R. Elisabeth Cornwell, Richard Rumary, Alan Heesom, Ian McAlpine, Michael Ottway, Howard Stringer, Anna Sander, Paula Kirby, Stephen Freer, Bart Voorzanger, Jennifer Jacquet, Lucy Wainwright, Bjorn Melander, Christer Sturmark, Greg Stikeleather, Ann-Kathrin Ehlers, Jan and Richard Gendall, Rand Russell.

TEXT ACKNOWLEDGEMENTS

Every effort has been made to trace the copyright holders, but any who have been overlooked are invited to get in touch with the publishers.

'To the Balliol Men Still in Africa' by Hilaire Belloc reprinted by permission of Peters Fraser & Dunlop (www.petersfraserdunlop.com) on behalf of the Estate of Hilaire Belloc.

Extract from *Iris Murdoch: A Life* by Peter J. Conradi © Peter J. Conradi, 2001, reprinted by permission of A. M. Heath & Co Ltd and W. W. Norton.

Extract from *The Autobiography of Bertrand Russell* by Bertrand Russell © 2009 The Bertrand Russell Peace Foundation, reprinted by permission of Taylor & Francis Books UK and The Bertrand Russell Peace Foundation Ltd.

Lyrics from 'A Song of Reproduction' reprinted by permission of the Estates of Michael Flanders & Donald Swann 2013. Any use of Flanders & Swann material, large or small, should be referred to the Estates at leonberger@donaldswann.co.uk.

Extract from 'Summoned by Bells' from *Collected Poems* by John Betjeman © 1955, 1958, 1962, 1964, 1968, 1970, 1979, 1981, 1982, 2001 reprinted by permission of John Murray (Publishers) and The Estate of John Betjeman.

Extract from 'A Hike on the Downs' from *Collected Poems* by John Betjeman © 1955, 1958, 1962, 1964, 1968, 1970, 1979, 1981, 1982, 2001 reprinted by permission of John Murray (Publishers) and The Estate of John Betjeman.

Extract from *The Loom of Years* by Alfred Noyes © 1902 reprinted by permission of The Society of Authors as the Literary Representative of the Estate of Alfred Noyes.

'Blue Suede Shoes' by Carl Lee Perkins © 1955, 1956 Hi Lo Music, Inc. © Renewed 1983, 1984 Carl Perkins Music, Inc. Administered by Wren Music Co., Division of MPL Music Publishing, Inc. All rights reserved. International copyright secured. Used by permission of Music Sales Limited.

Extract from *The Silent Traveller in Oxford* by Chiang Yee © 1944 Signal Books Ltd.

Extract from W. D. Hamilton, 'The Play by Nature', *Science* 196: 757 (1977), reprinted with permission from AAAS.

Extract from *Leda* by Aldous Huxley. Copyright © 1929 by Aldous Huxley. Reprinted by permission of Georges Borchardt, Inc., on behalf of the Aldous and Laura Huxley Trust. All rights reserved.

Extract from 'Genes and Memes' by John Maynard Smith, first published in *London Review of Books*, 4 February 1982.

Extract from 'Selective Neurone Death as a Possible Memory Mechanism' by Richard Dawkins, first published in *Nature* (Nature Publishing Group), 8 January 1971.

Extract from Richard Dawkins' Foreword to John Maynard Smith, *The Theory of Evolution* (Cambridge University Press, 1993).

Extracts from Preface, chapter 1 and chapter 13 of *The Selfish Gene* by Richard Dawkins (1976). Reprinted by permission of Oxford University Press.

PICTURE ACKNOWLEDGEMENTS

All photos come from the Dawkins family collection (thanks to Sarah Kettlewell) except where otherwise acknowledged. Every effort has been made to trace copyright holders, but any who have been overlooked are invited to get in touch with the publishers.

Images in the text

p. 181: *Cerura vinula*: photo courtesy N. Tinbergen.

Illustration sections

Section one

St Mary's Church, Chipping Norton: photo courtesy Nicholas Kettlewell.

Clinton Edward Dawkins (1880), Clinton George Evelyn Dawkins (1902), Clinton John Dawkins (1934), Arthur Francis 'Bill' Dawkins (1935/6): photos courtesy Balliol College, Oxford.

Section two

Emperor Swallowtail (*Papilio ophidicephalus*): © Ingo Arendt/Minden Pictures/Corbis.

Section three

The Great Hall, Oundle School, Northamptonshire: © Graham Oliver/ Alamy; Ioan Thomas, 1968: Oundle School Archive.

Niko Tinbergen painting hens' eggs to resemble gulls' eggs, c.1964: Time & Life Pictures/Getty Images; Mike Cullen, 1979: Monash University Archives, photo Hervé Alleaume; the Surrey Puma hunt: photo courtesy Virginia Hopkinson; People's Park demonstrators and the National Guard, Berkeley, 19 May 1969: © Bettmann/Corbis; punting in Oxford: photo courtesy Lary Shaffer; Peter Medawar at University College, 26 November 1960: Getty Images.

RD and Ted Burk, November 1976: Time & Life Pictures/Getty Images; Danny Lehrman and Niko Tinbergen: photo courtesy Professor Colin Beer; Niko Tinbergen filming: courtesy Lary Shaffer.

William D. Hamilton and Robert Trivers, Harvard, 1978: photo courtesy Sarah B. Hrdy; Michael Rodgers: photo courtesy Nigel Parry; RD and George C. Williams: photo by Rae Silver courtesy John Brockman; John Maynard Smith: Corbin O'Grady Studio/Science Photo Library; *The Selfish Gene*: courtesy Keith Cullen.

INDEX

Richard Dawkins was first catapulted to fame with his iconic work *The Selfish Gene*, which he followed with a string of best-selling books: *The Extended Phenotype, The Blind Watchmaker, River Out of Eden, Climbing Mount Improbable, Unweaving the Rainbow, The Ancestor's Tale, The God Delusion, The Greatest Show on Earth, The Magic of Reality*, and a collection of his shorter writings, *A Devil's Chaplain*.

Dawkins is a Fellow of both the Royal Society and the Royal Society of Literature. He is the recipient of numerous honours and awards, including the Royal Society of Literature Award (1987), the Michael Faraday Award of the Royal Society (1990), the International Cosmos Prize for Achievement in Human Science (1997), the Kistler Prize (2001), the Shakespeare Prize (2005), the Lewis Thomas Prize for Writing about Science (2006), the Galaxy British Book Awards Author of the Year Award (2007), the Deschner Prize (2007) and the Nierenberg Prize for Science in the Public Interest (2009). He retired from his position as the Charles Simonyi Professor for the Public Understanding of Science at Oxford University in 2008 and remains a fellow of New College.

In 2012, scientists studying fish in Sri Lanka created *Dawkinsia* as a new genus name, in recognition of his contribution to the public understanding of evolutionary science. In the same year, Richard Dawkins appeared in the BBC Four television series *Beautiful Minds*, revealing how he came to write *The Selfish Gene* and speaking about some of the events covered in this memoir.

In 2013, Dawkins was voted the world's top thinker in *Prospect* magazine's poll of 10,000 readers from over 100 countries.